Technologies for Engineering Manufacturing Systems Control in Closed Loop

Dissertation

zur Erlangung des Doktorgrades der Ingenieurwissenschaften (Dr.-Ing.)

des Zentrums für Ingenieurwissenschaften

der Martin-Luther-Universität
Halle-Wittenberg

vorgelegt

von Herrn Dipl.-Ing. Sebastian Preuße
geboren am 24.04.1983 in Bernburg

Gutachter

1. Prof. Dr.-Ing. Hans-Michael Hanisch
2. Prof. Dr. Valeriy Vyatkin

Halle (Saale), 28. November 2013

Reihe: Hallenser Schriften zur Automatisierungstechnik
herausgegeben von:
Prof. Dr. Hans-Michael Hanisch
Lehrstuhl für Automatisierungstechnik
Martin-Luther-Universität Halle-Wittenberg
Theodor-Lieser-Str. 5
06120 Halle/Saale

email: Hans-Michael.Hanisch@informatik.uni-halle.de

Bibliographic information published by the Deutsche Nationalbibliothek

The Deutsche Nationalbibliothek lists this publication in the Deutsche Nationalbibliografie; detailed bibliographic data are available in the Internet at http://dnb.d-nb.de

(Hallenser Schriften zur Automatisierungstechnik; 10)

ISBN 978-3-8325-3600-8

Logos Verlag Berlin GmbH
Comeniushof, Gubener Str. 47,
10243 Berlin
Tel.: +49 030 42 85 10 90
Fax: +49 030 42 85 10 92
INTERNET: http://www.logos-verlag.de

Vorwort

Diese Arbeit entstand während meiner wissenschaftlichen Tätigkeit am Lehrstuhl für Automatisierungstechnik an der Martin-Luther-Universität Halle-Wittenberg. Maßgeblich inspiriert wurde sie durch die Bearbeitung des Forschungsverbundprojektes *On-the-fly-Migration und Sofortinbetriebnahme von automatisierten Systemen*, welches durch das Bundesministerium für Wirtschaft und Technologie gefördert wurde.

Ganz besonders danken möchte ich meinem Doktorvater und ersten Gutachter, Herrn Prof. Dr.-Ing. Hans-Michael Hanisch. Er hat meine wissenschaftliche Laufbahn bereits zur Zeit meines Studiums mit großem Engagement gefördert und mein Interesse an der Automatisierungstechnik geweckt. Als hilfswissenschaftlicher Mitarbeiter, dann als Diplomand und schließlich als wissenschaftlicher Mitarbeiter und Doktorand wurden mir stets alle notwendigen Mittel und Freiheiten zur Verfügung gestellt, um meine Forschungsarbeit weiter führen zu können. Die regelmäßigen Diskussionen mit Hans-Michael waren enorm hilfreich, um mich thematisch zu fokussieren. Er wendete vor allem für die Strukturierung des Manuskripts sehr viel Energie und Zeit auf, wofür ich ihm sehr dankbar bin. Darüber hinaus schätzte ich die vielen persönlichen Treffen und die Gewissheit, dass er immer ein offenes Ohr für mich hatte.

Meinem zweiten Gutachter, Herrn Prof. Dr. Sc. Valeriy Vyatkin, danke ich für die häufigen - meist zufälligen - Diskussionen auf Konferenzen, die mir stets kritische Hinweise zu Schwachstellen und zu Innovationspotentialen gaben und nicht selten Ansätze komplett in Frage stellten. Diese Erkenntnisse trugen entscheidend zur Arbeit bei.

Des Weiteren möchte ich Herrn Prof. Dr. Wolf Zimmermann für die vielen Hinweise zum formalen Teil der Arbeit danken. Hierdurch konnte ich die Qualität der Definitionen sowie der entwickelten Formalismen signifikant verbessern. Seine Ideen zum Ausbau verschiedener Ansätze haben mir Innovationspotentiale aufgezeigt und Vorschläge für zukünftige Arbeiten geliefert. Ferner fühlte ich mich in seiner Arbeitsgruppe stets willkommen, was ein angenehmes Arbeiten ermöglichte.

Meinen Kollegen am Institut für Informatik - insbesondere am Lehrstuhl für Automatisierungstechnik - danke ich für die erfolgreiche Zusammenarbeit und das eine oder andere persönliche Gespräch. Im Besonderen danke ich Frau Dr. Roswitha Picht, die meine Wissenslücken konsequent mit viel Geduld, Zeitaufwand und endlosem Optimismus füllte und mir für die Fertigstellung der Arbeit den Rücken frei hielt.

Schließlich danke ich meiner Familie und meinen vielen Freunden - insbesondere Herrn Ringo Kämmler - für das große Interesse am Entstehungsprozess der Arbeit sowie für die ab und an notwendige Zerstreuung.

Abstract

Today's world is full of smart gadgets, embedded devices, and computer systems. To use this equipment, software is implemented, which processes data and helps to make decisions, to control the distribution of goods, or to monitor processes - to mention just a very tiny subset. The correctness of this software is crucial for a failure-free but also safe operation of the corresponding systems.
Concentrating on the topic of software correctness, this thesis is particularly focused on manufacturing control systems and their related characteristics. Thereby, the closed loop of the controller and the controlled plant is regarded, which clearly distinguishes this work from the majority of other scientific contributions. It claims to serve engineering demands for a systematic but also comprehensible enrichment of methodologies that are already applied in practice. For this, the mayor contribution of this work is a framework, which combines different solutions from academia to be applied in the plant engineering environment. These formal approaches concern the (semi-) automatic plant and controller model generation, the domain-specific description of the system's behavior, the test case generation, the improved simulation of the overall process, and last not least, the verification by means of model checking.
The framework supports plant engineering from the beginning on, which encloses system design, implementation, documentation, maintenance, as well as testing and verification. Thereby, it is integrated to the engineering workflow and applies data, which usually is available already. Because of this, it does not open a completely new branch for system's analysis, but it fits in the conventional engineering process. The framework does not depend on a specific controller hardware vendor as there are no restrictions according to the fields of application. However, it is tailored - but not limited - to IEC 61131-3-conform software implementations. To increase the user-acceptance, it reveals software failures within the control code and gives advices how to fix these. Most steps are automated to prevent the user from dull theory. However, the control algorithms are not modified by the applied software tools, but the software engineer retains full control over the process.
Properly applied, the correctness of the control software is formally proven by the analyzing framework according to the specified requirements. Although, additional efforts have to be made, the costs for error tracing and correction while starting up or even while running the technical plant are economized.

Kurzdarstellung

In vielen Bereichen des täglichen Lebens sind elektronische Hilfssysteme, intelligente Steuergeräte oder Computer allgemein feste Bestandteile. Um sie möglichst flexibel an die gegebenen Rahmenbedingungen anpassen zu können, sind sie programmierbar. Die Korrektheit der dazu notwendigen Software ist entscheidend für die Sicherheit der Gerätschaften sowie für die optimale Bewältigung der gestellten Aufgaben.
Ziel dieser Arbeit ist es, die Korrektheit der Steuerungssoftware für Produktionsanlagen mit deren spezifischen Eigenschaften zu gewährleisten. Dazu wird der geschlossene Kreis aus Steuerung und technischer Anlage betrachtet, was diese Arbeit gegenüber den meisten anderen wissenschaftlichen Beiträgen deutlich abgrenzt. Sie erhebt den Anspruch, konventionelle ingenieurmäßige Arbeitsabläufe um systematische und verständliche Methoden zu erweitern. Aus diesem Grunde ist das Hauptanliegen dieser Arbeit, verschiedene theoretisch fundierte Lösungsansätze auf praktische Problemstellungen anzuwenden und somit ein Rahmenwerk zu schaffen, welches die formale Analyse eines technischen Systems praktikabel macht. Kernpunkte sind dabei die (halb)automatische Generierung von formalen Anlagen- und Steuerungsmodellen, die domänenspezifische Spezifikation des Anlagenverhaltens, die Generierung von Testfällen, die erweiterte Simulation des technischen Prozesses sowie die Verifizierung des Steuerungsprogramms mit Hilfe der Modellprüfung.
Das Rahmenwerk begleitet den Anlagenentwicklungsprozess von Anfang an und unterstützt den Entwurf, die Umsetzung, die Dokumentation sowie die Wartung und ermöglicht schließlich das formale Testen bzw. Verifizieren. Dazu wird es in die bereits bestehenden Arbeitsabläufe des Ingenieuralltages integriert und verwendet vorhandene Daten, um Modelle und Anwendungsfälle abzuleiten. Es ist herstellerunabhängig und besitzt keine Einschränkungen bezüglich der verwendeten Steuerungshardware. Im Kontext dieser Arbeit werden IEC 61131 konforme Implementierungen betrachtet. Die Anwendung ist jedoch nicht auf diese beschränkt.
Fehler, die bei der Analyse aufgedeckt werden, werden aufbereitet und visualisiert. Die eigentliche Anpassung des Steuerungsprogramms wird ausschließlich durch den Softwareingenieur vorgenommen. Somit entscheidet dieser manuell über gegebenenfalls notwendige Änderungen, was die Akzeptanz des vorgestellten Ansatzes steigert. Der Korrektheitsnachweis bezüglich einer Verhaltensspezifikation wird für alle möglichen Anlagenzustände erbracht. Der zusätzliche Aufwand für diese formale Analyse amortisiert sich spätestens bei der Inbetriebnahme bzw. im laufenden Betrieb, da sowohl die kostenintensive Fehlersuche als auch die -korrektur entfallen.

Contents

List of Figures

List of Tables

List of Abbreviations

Abbreviation	Meaning
*.cis	CAD import script
*.dxf	CAD drawing interchange file format
$<e_1,e_2>$	path between two edges e_1 and e_2 of a graph
AutomationML	Automation Markup Language
CAD	Computer Aided Design
CoDeSys	Controller Development System
CTL	Computation Tree Logic
DFG	Deutsche Forschungsgemeinschaft (German Research Foundation)
${}_{D}$TNCE	Discretely-Timed Net Condition/Event
${}_{D}$TNCEM	Discretely-Timed Net Condition/Event Module
${}_{D}$TNCES	Discretely-Timed Net Condition/Event System
EBNF	Extended Backus-Naur Form
eCTL	extended Computation Tree Logic
EMC	Extended Model Checker
EnAS	Energy-Autarkic Actuator and Sensor System
FBD	Function Block Diagram
Flex	Fast Lexical Analyzer
GUI	graphical user interface
HiL	Hardware-in-the-Loop
I/O	Input/Output
IED	Incontrol Enterprise Dynamics
IL	Instruction List
LD	Ladder Diagram
MDE	Model Driven Engineering
$\mathbb{N}$	Set of natural numbers (excluding zero).
$\mathbb{N}_0$	Set of natural numbers (including zero).
NCEM	Net Condition/Event Module
NCES	Net Condition/Event System
OMSIS	On-The-Fly-Migration and Instant-Start-Up of Automated Systems
PandIX	Piping and Instrumentation Diagram Exchange
PLC	Programmable Logic Controller
SESA	Signal-Net System Analyzer

SFC .... Sequence Function Chart
SIL .... Safety Integrity Level
SiL .... Software-in-the-Loop
SNS .... Signal-Net System
SOTL .... Safety-Oriented Technical Language
ST .... Structured Text
STD .... Symbolic Timing Diagram
SVM .... Symbolic Model Verifier
TCTL .... Timed Computation Tree Logic
TIG .... Transition-Invariants Graph
TMoC .... TNCES Model Checker
VDE .... Verification-Driven Engineering

1. Introduction

Automation technology is a quickly-emerging engineering domain since the degree of automation is increasing in almost all areas of life like traffic, communication, manufacturing, chemical industry or habitation. The world is full of micro controllers, embedded systems, and personal computers that serve humans either observably or hidden in countless situations of daily life. Besides the failure-free operation of hardware, that - amongst others - is affected by physical laws, software is the most crucial factor for a save and target-oriented interaction with all these small assistants. According to Moore's Law [1], the computation power of micro controllers doubles almost every 2 years, and because of this, embedded devices have become able to process complex software that executes more and more tasks. However, the methods of software implementation have not changed remarkably because usually, it is still written and tested manually, and for this, the probability of software errors increases with its complexity.

Software failures have been occurring since software has been applied to code programmable devices. There are countless examples for bugs that actually have been filling books. For example, on New Year's Eve 1999/2000 the world was holding its breath expecting computer systems all over the planet to crash because of the year 2000 problem, which is also known as the millennium bug [Ain96]. Fortunately, nothing serious happened. Probably, the most expensive results of failures are to be found in history of space flight. On July 22nd, 1962, the Mariner 1 spacecraft[2] was blasted by the range safety officer because it veered off course. Analysts found out that an overbar ("$^-$") within the handwritten notes of mathematic calculations was not implemented in the control program. For this, the event went down in history as: "The most expensive hyphen in history." The failed maiden flight of the Ariane 5 spacecraft[3] on June 4th, 1996, is a further prominent example for software failures. Engineers migrated parts of the control software of the predecessor Ariane 4, which worked without reproach, but they were not aware of the changed physical parameters as Ariane 5 was bigger and accelerated more powerfully. Consequently, the speed, which was measured after launch in terms of a 64 bit float point number, caused a memory overflow and overwrote navigation data while converting it to a 16 bit fix point number. The com-

---

[1] Moore's Law. URL, http://www.intel.com/technology/mooreslaw/, November 2013.

[2] NASA: Mariner 1. URL, http://nssdc.gsfc.nasa.gov/nmc/spacecraftDisplay.do?id=MARIN1, November 2013.

[3] ESA: Ariane 501. URL, http://www.esa.int/esaCP/Pr_33_1996_p_EN.html, November 2013.

puter interpreted the corrupted data as course deviation and tried to compensate it. Finally, this resulted in a crash of the spacecraft. On the roads, software bugs are to be found as well. In May 2005, Toyota had to call back its Prius cars because a software bug caused the engine to shut down on highway.

The world is full of software and consequently of software bugs. Finding and correcting them is a great challenge. Even very experienced software vendors sell products containing weak points. Nearly every PC user is reminded of that while switching on its computer and being prompted for downloading new updates.

As this conclusion also holds for the plant engineering domain, the contribution of this thesis is to support the software engineering for industrial controllers. For this, it combines practical problems of daily plant engineering and theoretical solutions of computer science to assist the development of meaningful control software. The main focus of the work is to provide an integrated approach that makes use of existing and established procedures, and to enrich them with formal analyzing possibilities. To stay as close as possible to practical engineering, the theory is hidden in background and provided with intuitive and domain-specific front-ends. Because of this, the work does not present further theory and formalisms that have to be studied and learned, but it is tailored to the demands of professionals by extending their conventional software engineering approaches with a framework, which can be fully integrated into their accustomed workflows.

The work is motivated by challenges of control software implementation and therefore has a practical reference to the domain of plant engineering. The first section of the introduction is focused on existing problems of controller implementation. Subsequently, the framework of this thesis is presented, and afterwards, the structure of the work is depicted.

1.1. Motivation

Software implementation expenses for manufacturing control systems, both in time and money, have gained a level, where software is no longer just a byproduct but a real cost factor for plant engineering. This shows up all the more since plant engineering is a very integrated task, which is performed in parallel and in team work between engineers of different technical domains. Control software is implemented while the real plant is constructed and due to time-to-market pressure, its correctness is extensively tested first while starting up the plant. Consequently, the start-up might be delayed by software failures, which have to be debugged. Beyond this, there can exist errors that happen rarely but are the more critical ones because they occur seemingly randomly, and discovering and correcting their origin can be cost-intensive. Conventionally, plant engineers apply manual testing procedures to approach the problem of incorrect control software. However, those methods have clearly reached their limits.

Since the degree of automation is increasing continuously, it is not possible anymore to cover every failure scenario and to exclude critical behavior with manual methods. Gradually, a paradigm shift is recognizable in the plant engineering domain since Model-Driven Engineering (MDE) [GDD09] approaches are incrementally applied to model continuous and discrete processes. Doing so, new plants are modeled and simulated long before they are constructed, and the control software is not implemented writing innumerable code lines anymore, but it is derived from software models automatically. While this procedure was an exclusive topic for academia a few years ago, it has now found its way into the industrial practice as well. However, plant and controller models offer far more possibilities than just simulating the controlled process. If they are based on formal models, they further facilitate the verification of control software. While simulating, certain test cases are executed to check for the correct plant operation. This procedure is not complete since rare but critical scenarios might be overseen, which can rather affect the system's safety. Because of this, simulation is a powerful tool to validate the correct function of the control software, but it does not provide a final assertion about its correctness. In contrast, verification considers every possible state of the controlled system, and for this, it provides a machine-based proof of correctness. Doing so, the MDE approach is extended to the Verification-Driven Engineering (VDE) approach. Applying it to the plant engineering domain is an innovation because to the best of the author's knowledge, there is no integrated VDE approach so far that is applied in practice. This assumption is supported by the contribution of [Joh07], which gives a summary of applications of formal technologies in manufacturing systems control. The author states that *to date, very few opportunities for formal verification have been realized [...].*
One remarkable reason is that formal methods usually come along with complex theory. Consequently, the motivation of this thesis is not to provide further theory and formalisms, but it shall be tailored to the demands of plant engineers. It applies proven methods from computer science to extend the conventional analyzing approaches for automated systems. For this, it proposes methodologies how to integrate the theoretical solutions from academia to solve the current challenges of practice.
While writing this thesis, several times the author was asked by economists, why additional efforts should be spent to establish a verification framework that obviously requires expertise and man-power. To build a bridge to the world of economists and to evaluate the cost-benefit factor with concrete numbers, one is in need of operating figures like energy consumption, avoidance of downtimes or prevention of accidents. As this is possible only with statistics and reliable field tests, these concrete numbers are not provided in this thesis because corresponding experiments producing dependable data were not made. Moreover, usually companies are by no means interested to provide such data for publication. However, in ideal case the framework verifies the control software and prevents engineers from randomly searching for bugs in rather complex systems. The logical conclusion can be drawn that this benefit is going to outweigh the additional costs the bigger systems get. For this, the question should

not be whether the application of automated verification technologies is meaningful in practice, but how such methodologies can effectively be integrated into the engineering work-flow.

A further point that is addressed in this work is migration. It has gained in importance in the last years since process control systems that are older than thirty years - which is not uncommon due to the introduction of such systems in the 1960s - do not meet increasing demands according to computation power and signal processing anymore. This shows up when a plant is going to be expanded and the outdated process control system has to take over more and more control functionalities. Quickly, engineers are faced with problems because of a limited signal interface or antiquated programming technologies. Beyond this, the supply with spare parts as well as maintenance support of manufacturers are discontinued the older the systems get. While process control systems were first applied only to measure a few signals and to visualize them in measuring stations, today they carry the whole technical process, react to failures, and have nearly substituted human beings in plants. To come along with the rising demand for computation power and for efficient process signal handling, new systems have to take over the functionalities of aged process control systems. However, this modernization has to happen without disturbing the technical process itself because due to economical reasons, producing plants cannot be shut down completely. Consequently, the migration progress is complicated since debugging of control software and reorganizing of outdated control structures have to be performed with minimum influence on the real plant. Therefore, this thesis approaches this problem and proposes a framework to model the technical process and to validate and verify the control software. Doing so, plant downtimes are minimized, and furthermore, the costs for the actual control hardware and software modernization can be reduced.

Summing up, the crucial point is the feasibility of control software that must be ensured even before interconnecting controller and plant. This is important for safety reasons to avoid dangerous situations as well as for performance considerations to ensure that the controller is able to run the software and consequently the plant under certain specified conditions. The complete approach is summarized in a framework that claims to be applicable in practice and has the capability to be integrated into the daily life of engineering. In the subsequent section, this framework is presented.

1.2. Framework of this Thesis

As motivated in the previous section, the contribution of this thesis is an integrated framework for the verification of manufacturing control systems. For this, it combines heterogeneous expertise of research fields from engineering and computer science. In principle, the central theme is given by the following three requirements that have been the guideline for evaluating and developing approaches:

- Provide a language that engineers accept as “natural”, which means that domain-specific description technologies shall be melt into novel approaches to gain maximal user acceptance.
- Make formal analysis as automatic as possible, which means that the user must not be overcharged with complex theory, but intuitive software solutions with familiar interfaces shall be provided. Beyond this, user interaction should be reduced to a minimum to avoid operator errors.
- Provide good feedback, which means that the results of formal analysis shall be presented in a comprehensible way.

The technologies for analysis are based on a Hardware-in-the-Loop (HiL) approach on the one hand, and on a Software-in-the-Loop (SiL) approach on the other hand. Both are addressed to particular fields of application. While performing HiL tests, the control software runs on the target controller and is analyzed for “real” conditions. Besides the logical correctness of the control code, especially the impacts of timing aspects like communication delays or time functions within the control code are considered. On the other hand, SiL tests are performed without determining a specific hardware. For this, they are executed independently from the particular controller hardware.

This thesis' topic is especially influenced by the industrial joint research project *On-The-Fly-Migration and Instant-Start-Up of Automated Systems*[4] (OMSIS), which has been processed by the author in collaboration with others. The project was referred to practical demands, whereas interdisciplinary expert knowledge from industrial partners flew in. Because of this, the focus was on a target-oriented and easy application in the context of practical engineering.

The part of the author's job was to establish the framework for formal modeling, specification, and verification. It is the opinion of the author that such a framework only will be applicable in a meaningful context if the closed loop of the controller and the controlled process is considered. Of course, open loop technologies have their approved application areas in computer science, but in the (discrete) control domain, considering the real plant or at least its model is inevitable to develop the control software. This perception is no innovation because the continuous control field depends on plant models from the very first day since continuous processes have been automated. In other engineering domains, like for example aircraft or automotive industry, closed-loop technologies are already employed. The expenses are justified because the quantity of produced vehicles is comparatively high and critical software parts have to be 100-percently safe. Plant engineering is slightly different because every plant is unique and the established methodologies are rather conservative. Nevertheless, the author claims that it will benefit from verification as well. Thereby, the wheel

[4]OMSIS project website. URL, http://iai8292.inf.tu-dresden.de/omsis/de/index.html, November 2013.

should not be reinvented in this work, but established expertise shall be adapted to the manufacturing control field. Summing up, the author would like to emphasize that the topic of this thesis is clearly distinguished to any other work, which performs open-loop simulation and verification. On the other hand, it is interdisciplinary and joins different theoretical approaches to be applied in practice. To give an overview, a survey of the framework is described subsequently. The details are provided in the following chapters.
Usually, the starting point in plant engineering is a technical plant specification, which describes the project parameters. Based on it, the plant is designed using a Computer Aided Design (CAD) tool to plan the necessary construction steps and material requisitions. This CAD drawing, which is shown in the upper left side of Figure 1.1, is the starting point for the presented framework since it is further employed to model the plant. It contains the static plant structure but not its dynamics. In the scope of this thesis, the commercial simulation software *Incontrol*[5] *Enterprise Dynamcis* is applied. The software imports the structural CAD data ① and supports defining the moving plant parts to develop the plant simulation model.This plant simulation is suitable to check the physical plant behavior to exclude collisions of moving parts, or to check whether the specified production scenarios can be executed. To do the latter, the controller is interconnected with it ② to run a HiL or a SiL Simulation. For the HiL Simulation the control code is executed on the target controller. In contrast, this code runs in a simulated environment on a PC while performing the SiL Simulation.
A remarkable disadvantage of simulation is the restriction to certain test cases. Even though these test cases cover many failure scenarios, a complete consideration of all possible plant states cannot be ensured. This is possible only by performing verification. For this, the plant simulation model is translated to a formal plant model ③. To do so, *Discretely-Timed Net Condition/Event Systems* ($_D$TNCES) are applied as modeling formalism. The translation process is widely automated and supported by a software tool. It analyzes the model data and maps the different plant modules of the simulation to their corresponding formal plant $_D$TNCE Modules. The modules of the formal plant model are manually created for each plant part. Once generated, they are reusable for other plants, and combined in a library of frequently-used plant modules to support the translation process. Like for the simulation, there are two possibilities to verify the controlled system, namely HiL and SiL Verification. For HiL Verification, the controller is connected to the formal plant model, and a reachability analysis is performed ④. In contrast, for SiL Verification the control code is translated to a formal controller model. This controller model is composed with the formal plant model to a closed-loop system model, and again, a reachability analysis is performed ④. The computed HiL dynamic graph contains all controllable plant model states plus the in- and output states of the controller hardware. On the other hand, the SiL dynamic graph includes every possible state of the closed loop of plant and controller model.

[5]Incontrol website. URL, http://www.incontrolsim.com, November 2013.

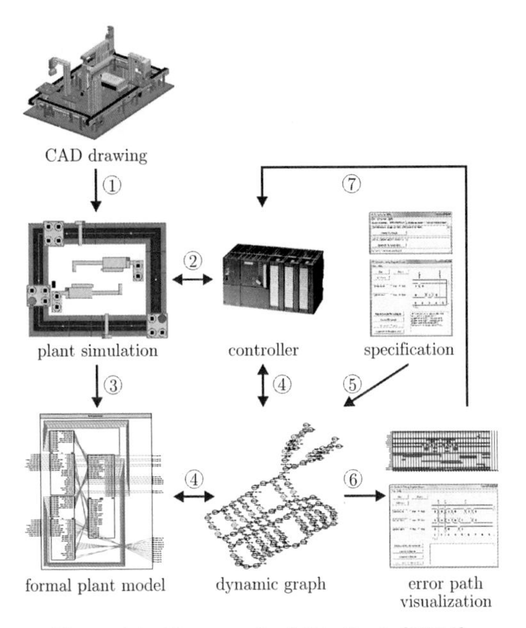

Figure 1.1.: Framework of this thesis [PH11].

If it is complete or a certain final state is reached, a formal specification of behavior is applied to it ⑤ to do model checking [CGP00]. The specification is expressed in a temporal logic and involves functional and non-functional properties. If a property is not fulfilled, a counterexample is generated that shows the trajectory through the dynamic graph from the initial to the error state. This error path is visualized ⑥ so that the control code can be fixed ⑦ if failures have been detected. The process is iteratively performed until the specification is fulfilled.
With reference to the three requirements, which are mentioned in the beginning of this section, the framework is tailored to verify the correctness of the control software according to a specification of behavior. This specification is developed with domain-specific description techniques. The formal plant models are generated semi-automatically so that the engineer, who applies the framework, does not have to be familiar with formal modeling in detail. However, the approach will be applicable only if meaningful plant models are provided. Hence, it is mandatory that the CAD data (that is the static model) and consequently the plant simulation (that is the dynamic model) are correct. This point is crucial because the success of applying MDE and VDE approaches depends on models that are accurate enough. For this, as many steps

as possible have been automated so that sources of errors are eliminated. However, the framework provides information, where errors in the control code can be found, but it does not generate control code. Therefore, it is independent from a particular hardware vendor, and it is applicable for control applications following the international standard IEC 61131-3 as well as those ones following the IEC 61499-1.
To the best of the author's knowledge, there exists no corresponding approach joining the different disciplines like that one presented in this work. Related contributions of other groups, which correspond to the particular work packages of this thesis, are presented in the respective areas of the different chapters. In the next section, the structure of this thesis is described.

1.3. Structure of this Thesis

The previous section provides an overview of the framework of this thesis. The subsequent chapters go into detail, and therefore, the structure of this work is given as follows. Chapter 2 lists theoretical background to the formalisms and abstract concepts, which are applied in the thesis. Afterwards, Chapter 3 gives attention to the formal modeling of plants and controllers. Beyond this, concepts are discussed to automatically generate formal models out of informal data. Chapter 4 presents two description languages, which have been picked up and developed further in this thesis to create a formal behavior specification with domain-specific technologies. Subsequently, Chapter 5 applies all the approaches of this thesis to analyze the behavior of the closed-loop system. To do so, the two technologies of simulation and verification are deployed. In addition, a case study is performed to investigate the practical applicability of the framework. Finally, the work is concluded in Chapter 6.
The appendix provides further information in terms of graphics and algorithms. Relationships are established in the related chapters accordingly.

2. Basic Principles

The framework of this thesis fits in the closed-loop technology scheme in Figure 2.1. This scheme has been developed since 15 years [Han98] in the work group the author belongs to. The main focus is to improve the usual way of implementing the control software for a manufacturing system. The basis for an engineering project is the technical specification. The ordinary practice, depicted by a dashed line in Figure 2.1, is to interpret this specification and to implement the control code. This work highly depends on the experience of software engineers, and is done without support of formal technologies, but with established methodologies that have been improved over years in each company. Although these methodologies indeed work, especially the correctness of control software is a risky factor while scheduling the overall engineering process because the time, which is necessary to find and correct bugs, is hardly projectable.
The scheme in Figure 2.1 therefore proposes an improvement of conventional and antiquated control software implementation technologies in manufacturing systems' engineering by integrating formal analyzing methodologies to this process. Properly applied, the correctness of control software is proven and the costs for error tracking and debugging after starting up the manufacturing plant are saved.
This thesis claims to provide the concepts for the integration. For this, already existing data is applied to keep efforts at a minimum and to support the engineering processes instead of overcharging them with complex theory. Particularly, the field of *simulation and model checking* is in focus. The structure of this thesis follows the organization of the scheme. First of all the actual *plant* is considered. It is discussed, how formal plant *modeling* is performed to have a basis for closed-loop analysis. Afterwards, the *controller* is taken into account. Challenges exist according to control code implementation and execution behavior. These are approached to *transform* control code to formal models. Then, the concern of deriving a formal *specification* of behavior is addressed to achieve a well-defined but also understandable description of requirements. The *composition* of plant and controller models forms the *closed-loop system model*, which is analyzed by means of *model checking*. The results are then *interpreted* to draw conclusions according to the correctness of the control software. The topic of *synthesis* is excluded because it comprises a self-contained research field. Therefore, the interested reader is referred to [Mis12], where synthesis is approached in detail.

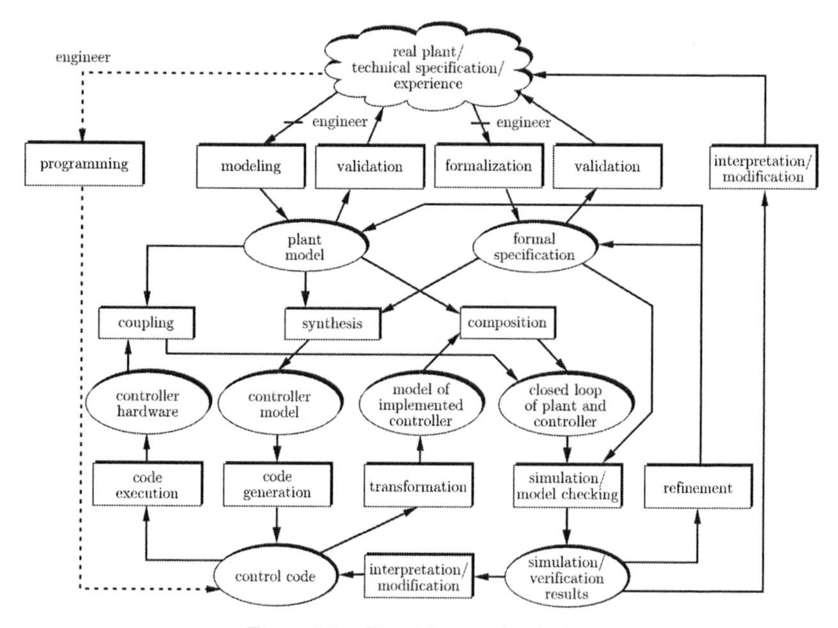

Figure 2.1.: Closed-loop technologies.

2.1. Technologies for studying Behavior

The terms simulation, verification, and validation are frequently mixed up and used in a mistakable context. For this, the application in this thesis shall be clarified in the following.
A simulation is the execution of a dynamic process of a system with the help of an executable model, to gain insight which can be transferred to reality; in a broadened meaning, a simulation is the preparation, execution and evaluation of targeted and focused experiments with a simulation-model [VDI96]. In the context of plant engineering, simulation means the application of test cases to a model of the regarded system, which represents the controller and/or the plant. The aim is to draw conclusions concerning the behavior of this real system. Actually, the specification of test cases is the weak point of this approach. It is hard to evaluate the completeness of simulation because each test case considers only one of the computation paths of a system. That means that tests can be incomplete and rare but critical behavior might be overseen. Consequently, undesired plant behavior could occur during runtime because it has not been specified and checked.

A formal approach to analyze systems' correctness is given by verification. It is the *confirmation, through the provision of objective evidence, that specified requirements have been fulfilled* [IEC08b]. In other words, a formal system model is checked against a formal specification of behavior. Doing so, the whole system state space is considered and assertions about system properties hold for the entire system. In the scope of this thesis, verification is performed as model checking. The approach highly depends on a meaningful and comprehensive specification. Furthermore, computing the whole state space can become a very complex task, which is commonly known as the state space explosion problem [LCL87]. This is one of the main reasons that verification is rarely applied for large-scale systems.
The behavior of a system is checked by performing validation. It *is the confirmation by examination and the provision of objective evidence that the particular requirements for a specific intended use are fulfilled.* [DIN05] Doing so, it is checked whether a system meets the requirements of practice. For this, external influences, such as user interaction, resource properties, environmental impacts, or production scenarios are considered. Validation can be performed by simulation because it evaluates whether test cases are appropriate for the specific use case.
All technologies rely on a meaningful system model. For this, deriving a model, which is as detailed as necessary and as abstract as possible, is one focus of this thesis. To study the behavior of systems, the preceding technologies are either applied as Hardware-in-the-Loop or as Software-in-the-Loop tests. In both cases, the environment, which is the plant, is given as a (formal) model. Performing HiL studies, the implemented control software runs on the target controller hardware and is interconnected with the plant model to establish the closed loop. Doing so, the control software can be analyzed under "real" conditions according to the level of detail of the plant model. In contrast, a model of the control code is abstracted to run SiL studies. This approach is of advantage if the execution environment, which is the controller, is not determined, yet. The specific advantages of HiL as well as of SiL tests are discussed later in this thesis.

2.2. Plants

Plant engineering covers the design, the construction, the start-up, and finally the maintenance of a technical system. This system can be a chemical production plant, a power station, a cogeneration plant, a storage power station, or a warehouse to mention just a very tiny subset. In the scope of this thesis, manufacturing systems are considered. These are *systems coordinated by a particular information model to support the execution and control of manufacturing processes involving the flow of information, material and energy in a manufacturing plant* [ISO09]. This view includes actuators and sensors as well as communication and power supply facilities, which

are necessary to handle raw materials and to provide products. The documentation of a manufacturing plant is done according to different views on the system. The ISO 15519-1 [ISO10] standard contains a summary of different document types such as piping and instrumentation diagrams, parts lists, or installations diagrams. Having regard to the plant model generation, especially the static CAD drawing is of importance for the framework presented in this thesis. The challenge is to extract all necessary information to establish a dynamic model, which is further applied for simulation and verification.
An unimpressive but crucial point is the later maintenance of a technical plant. For this, the documentation has to be consistent and changes in the system have to be captured in the documents. Unfortunately, this point is often neglected in practice. For this, migration of aged technical plants is going to be a rising challenge for plant engineering.

2.3. Controllers

A Programmable Logic Controller (PLC) is a means that is used to control and to automate industrial processes. It is connected to a plant or a physical process in general via input and output units. Its purpose is to control the state of the outputs depending on the states of the inputs and its internal state. Mainly, there are two international standards that define the implementation and execution of control software, namely the IEC 61131-3 [IEC03] and the IEC 61499-1 [IEC05]. Both are introduced in the following.

2.3.1. IEC 61131-3

The IEC 61131-3 is currently the most-applied standard to implement control software. It became widely used in automation industry since it has first been published in 1993. The IEC 61131-3 standard specifies a cyclic execution of the control software as illustrated in Figure 2.2. The control software can consist of several programs. Each one is assigned to a task that is either executed at defined times or runs as fast as possible. Despite the execution behavior and the program structure, the IEC 61131-3 defines two textual programming languages, namely Instruction List (IL) and Structured Text (ST) as well as three graphical languages, namely Function Block Diagrams (FBD), Sequence Function Charts (SFC), and Ladder Diagrams (LD).
However, syntax and semantics of these languages are not strictly defined, which leads to incompatibility of control software among different vendors. Due to the reason of the ambiguous standard, many hardware vendors have developed their own software engineering environments and regardless of the consistent functionality, the implemented control software is usually executable only on the corresponding hardware.

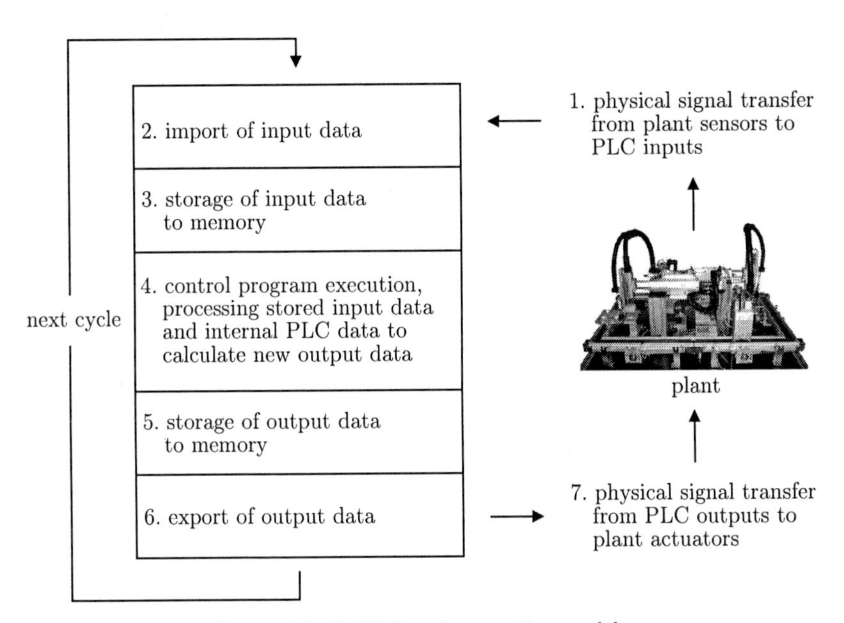

Figure 2.2.: 61131-3 cyclic execution model.

The *Controller Development System* (CoDeSys) is a license-free engineering environment of the *3S-Smart Software Solutions*[1] company that was founded in 1994. The first release also was published in 1994 and provides control software development capabilities as well as a hardware runtime environment. Over the years, the CoDeSys development system has been established as a quasi-standard among hardware vendors and until today, over 250 companies have joined the user group and adopted CoDeSys to implement engineering suites for their hardware.
The Technical Committee 6 of the PLCopen organization[2] approaches the problem of incompatibilities among engineering environments and defines an exchange format to transfer once developed function blocks, programs and PLC configurations between the engineering environments. An open source software tool already supporting this exchange format is *Beremiz* [3] [TBdS07], which is developed at the University of Porto. The authors provide an independent engineering environment to write PLC code in any of the five standardized languages. This code is then translated to a C++ program that can be compiled and executed independently from the platform. The authors

[1] 3S-Smart Software Solutions website. URL, http://www.3s-software.com, November 2013.
[2] PLCopen organization website. URL, http://www.plcopen.org, November 2013.
[3] Beremiz project website. URL, http://www.beremiz.org, November 2013.

claim that their software is applicable in academia as well as in industry. However, the attitude of automation industry towards new technologies is conservative. For this, Beremiz has to offer great efforts to be deployed.
The cyclic execution behavior of the IEC 61131-3 standard enables to calculate the time that is necessary to execute one cycle. This is especially important for real-time applications that require a defined minimum response time. However, the control software can be executed for many cycles without "recognizing" any changes of inputs or internal variables. Complex functions are unnecessarily called again and again and affect the maximal cycle time. Because of this, changes that might happen in the plant between two scans of the inputs are not recognized immediately. The IEC 61499-1 standard offers an alternative execution model and is introduced in the next section.

2.3.2. IEC 61499-1

The IEC 61499-1 [IEC05] is an industrial standard that defines an open architecture for the design of distributed control applications [PGH10]. The standard provides a generic model for distributed systems, which contains processes and communication networks as a basis for the distribution of applications on devices and their resources. Finally, it was standardized in 2005 and before that, it has been available as a Public Available Specification since 2000. In [Ger11], a comprehensive survey of verification of IEC 61499 control applications is given. As this thesis is tailored (but not limited) to the application of IEC 61131 implementations, the interested reader is referred to literature for more information on the IEC 61499.

2.4. System Models

The analyzing technologies of this thesis require a well-defined formalism, which supports system's modeling from the engineering point of view. In this context, $_D$TNCES [Ger11] are applied to model the behavior of the controlled system as well as of the controller itself. Thereby, the most important properties of $_D$TNCES are their:

- modularity and encapsulation of discrete behavior,
- hierarchical and component-based modeling framework,
- clearly-arranged graphical representation,
- possibility of being instantiated, parametrized and composed to larger systems,
- signal interfaces, and
- consideration of discrete time intervals.

The design of large systems out of modules is common to engineers. For this, $_D$TNCES are a very intuitive modeling formalism since modules describing basic functionalities are stepwise composed to a whole system model. The modularity enables the development of a library of frequently-used modules that can be used in other system modules again. This is of advantage especially while deriving formal plant models from informal simulation models. Beyond this, the formal basis of $_D$TNCES allows to apply formal analysis for example to calculate the dynamic graph - that is the set of all reachable states and state transitions of a $_D$TNCES - or to predict the behavior of a system under control. Originally, the timing concept and the means for dynamic graph computation were defined for (ordinary) Petri nets [HC95] and have been successfully applied in order to analyze and optimize different kinds of manufacturing systems. In the following, the formalism of $_D$TNCES is described.

2.4.1. Syntax

The syntax of $_D$TNCES is based on the known Petri nets [IEC04] consisting of places, transitions, and directed edges, which are called arcs as well. Ordinary arcs, or flow arcs respectively, are depicted by an arrow (⟶). These elements are extended by three kinds of signal arcs, namely condition arcs (—•), inhibitor arcs (—∘), and event arcs (⇢).

Definition 2.4.1 *($_D$TNCE Structure)*
$S = \{P, T, F, CN, I, EN, K, m, l, W_F, W_{CN}, W_I, ZF, sm, em\}$ is a $_D$TNCE Structure where:

- *P is the finite, non-empty set of places,*
- *T is the finite, non-empty set of transitions and $P \cap T = \emptyset$,*
- *$F \subseteq ((P \times T) \cup (T \times P))$ is the set of ordinary arcs,*
- *$CN \subseteq (P \times T)$ is the set of condition arcs,*
- *$I \subseteq (P \times T)$ is the set of inhibitor arcs and $F \cap CN \cap I = \emptyset$,*
- *$EN \subseteq (T \times T)$ is the irreflexive set of event arcs,*
- *$K : P \to \mathbb{N}$ assigns the capacity to each place,*
- *$m : P \to \mathbb{N}_0$ assigns the labeling to each place where $\forall p \in P : m(p) \leq K(p)$,*
- *$l : P \to \mathbb{N}_0$ assigns the local time to each place where:*
  $$l(p) = \begin{cases} 0 & if\ m(p) = 0 \\ n \in \mathbb{N}_0 & otherwise, \end{cases}$$
- *$W_F : F \to \mathbb{N}$ assigns the arc weight to each ordinary arc,*
- *$W_{CN} : CN \to \mathbb{N}$ assigns the arc weight to each condition arc,*

- $W_I : I \rightarrow \mathbb{N}$ *assigns the arc weight to each inhibitor arc,*
- $ZF : F \backslash (T \times P) \rightarrow \{[a;b] \, : \, a,b \in \mathbb{N}_0 \wedge a < b\}$ *assigns the time interval to each flow arc* (p,t) *where* $ZF(p,t) = \begin{cases} [0;\infty) & \text{if } \exists t' \in T \, : \, (t',t) \in EN, \\ [ZF_R; ZF_L] & \text{otherwise,} \end{cases}$
where ZF_R *is the retarding and* ZF_L *the limiting value of the time interval* $ZF(p,t)$*,*
- $sm : T \rightarrow \{i, s\}$ *assigns the firing mode to each transition* $t \in T$ *where* i *is the instantaneous and* s *is the spontaneous firing mode, and*
- $em : T \rightarrow \{\boxed{\wedge}, \boxed{\vee}\}$ *assigns the event mode to each transition* $t \in T$ *where* $\boxed{\wedge}$ *is the conjunction and* $\boxed{\vee}$ *the disjunction of incoming event signals.* □

Like in ordinary Petri nets, places are labeled with tokens in ${}_D$TNCES. The function $m(p)$ assigns a number of tokens to each place. In addition, the function $l(p)$ allocates a local time to each place, which indicates the age of its labeling. This local time will be zero if the place is not labeled. The time interval $ZF(p,t)$ specifies the earliest (ZF_R) and the latest (ZF_L) firing time of an enabled transition. A transition, which features an incoming event arc, must not have an incoming timed flow arc. In the graphical representation of ${}_D$TNCES, the time interval will only be shown if $ZF(p,t) \neq [0;\infty)$. A transition fires *spontaneously* or *instantaneously*, which is determined by the *firing mode* $sm(t)$. A spontaneous transition eventually fires as long as it is enabled. In contrast, an instantaneous transition immediately fires after it got enabled. In this thesis, only spontaneous transitions are applied to model the system's behavior. To handle multiple incoming event signals, the *event mode* $em(t)$ of a transition is considered. At a particular state transition, all incoming events have to fire in event mode $\boxed{\wedge}$. In contrast, any of them has to fire in event mode $\boxed{\vee}$. The standard event mode of a transition is $\boxed{\wedge}$, which is omitted in the graphical representation. Only, the event mode $\boxed{\vee}$ is represented explicitly.
To obtain modularization, the ${}_D$TNCE Structure is embedded into a module frame. This frame features a set of inputs and outputs, which are necessary to receive and to transmit condition and event information from and to the ${}_D$TNCE Structure and the module environment, respectively. The in- and output set is defined as follows.

Definition 2.4.2 *(In- and Output Set)*
An in- and output set is a tupel $\Phi = (C^{in}, E^{in}, C^{out}, E^{out})$ *where:*

- C^{in} *is the finite set of condition inputs,*
- E^{in} *is the finite set of event inputs,*
- C^{out} *is the finite set of condition outputs, and*
- E^{out} *is the finite set of event outputs.* □

The in- and outputs of the module frame are connected to the places and transitions of the ${}_D$TNCE Structure. Doing so, the in- and output structure is obtained.

Definition 2.4.3 *(In- and Output Structure)*
Let S be a ${}_D$TNCE Structure and Φ an in- and output set. Both form the in- and output structure $\Psi = (CN^{in}, W_{CN^{in}}, I^{in}, W_{I^{in}}, EN^{in}, CN^{out}, EN^{out})$ where:

- $CN^{in} \subseteq (C^{in} \times T)$ *is the set of condition input arcs,*
- $W_{CN^{in}} : CN^{in} \to \mathbb{N}$ *assigns the arc weight to each condition input arc,*
- $I^{in} \subseteq (C^{in} \times T)$ *is the set of inhibitor input arcs,*
- $W_{I^{in}} : I^{in} \to \mathbb{N}$ *assigns the arc weight to each inhibitor input arc,*
- $EN^{in} \subseteq (E^{in} \times T)$ *is the set of event input arcs where* $\forall (p,t) \in F$:
  $ZF(p,t) = \begin{cases} [0;\infty) & \text{if } \exists (e^{in}, t) \in EN^{in}, \\ [ZF_R; ZF_L] & \text{otherwise,} \end{cases}$
- $CN^{out} \subseteq (P \times C^{out})$ *is the set of condition output arcs where:*
  $\forall c^{out} \in C^{out} : \left|\{(p, c^{out}) \in CN^{out} : p \in P\}\right| \leq 1$, *and*
- $EN^{out} \subseteq (T \times E^{out})$ *is the set of event output arcs where:*
  $\forall e^{out} \in E^{out} : \left|\{(t, e^{out}) \in EN^{out} : t \in T\}\right| \leq 1$.

□

Informally, Definition 2.4.3 states that a transition with an incoming event signal only has incoming flow arcs without temporal constraints, which corresponds to Definition 2.4.1. Additionally, each condition and each event output is connected with at most one condition or one event arc, respectively. The in- and output structure is encapsulated into a Basic ${}_D$TNCE Module (${}_D$TNCEM).

Definition 2.4.4 *(Basic ${}_D$TNCE Module)*
Let S be a ${}_D$TNCE Structure, Φ an in- and output set, and Ψ an in- and output structure. Then, $\mathcal{M}_B = (S, \Phi, \Psi)$ is a Basic ${}_D$TNCEM. □

To develop hierarchical models, Basic ${}_D$TNCEM are composed to composite modules. Composite ${}_D$TNCEM consist either of Basic ${}_D$TNCEM or of other Composite ${}_D$TNCEM. Thereby, the modules are interconnected through event and condition arcs.

Definition 2.4.5 *(Composite ${}_D$TNCE Module and ${}_D$TNCE System)*
Composite ${}_D$TNCEM are defined inductively. Let $\{\mathcal{M}_1, \mathcal{M}_2, \ldots, \mathcal{M}_k\}$ be a finite and non-empty set of Basic or Composite ${}_D$TNCEM where $k \in \mathbb{N}$. Then, $\mathcal{M}_C = (\{\mathcal{M}_1, \mathcal{M}_2, \ldots, \mathcal{M}_k\}, \Phi, CK, EK)$ is a Composite ${}_D$TNCEM where:

- $\Phi = (C^{in}, E^{in}, C^{out}, E^{out})$ *is the in- and output set,*
- $CK \subseteq \Big(\bigcup_{j=1}^{k}(C^{in} \times C_j^{in})\Big) \cup \Big(\bigcup_{i=1}^{k}\big(\bigcup_{j=1}^{k}(C_i^{out} \times C_j^{in})\big)\Big) \cup \Big(\bigcup_{i=1}^{k}(C_i^{out} \times C^{out})\Big)$ *is the set of condition interconnections within* $\mathcal{M}_C$ *where:* $\forall c_s \in \big(C^{out} \cup (\bigcup_{j=1}^{k} C_j^{in})\big) : \Big|\big\{(c_q, c_s) \in CK \,:\, c_q \in \big(C^{in} \cup (\bigcup_{i=1}^{k} C_i^{out})\big)\big\}\Big| \leq 1,$
- $EK \subseteq \Big(\bigcup_{j=1}^{k}(E^{in} \times E_j^{in})\Big) \cup \Big(\bigcup_{i=1}^{k}\big(\bigcup_{j=1}^{k}(E_i^{out} \times E_j^{in})\big)\Big) \cup \Big(\bigcup_{i=1}^{k}(E_i^{out} \times E^{out})\Big)$ *is the set of event interconnections within* $\mathcal{M}_C$ *where:* $\forall e_s \in \big(E^{out} \cup (\bigcup_{j=1}^{k} E_j^{in})\big) : \Big|\big\{(e_q, e_s) \in EK \,:\, e_q \in \big(E^{in} \cup (\bigcup_{i=1}^{k} E_i^{out})\big)\big\}\Big| \leq 1.$

A Composite ${}_D$*TNCEM where* $\Phi = \emptyset$ *is a* ${}_D$*TNCES, denoted by:* $\mathcal{M}_S = (\{\mathcal{M}_1, \mathcal{M}_2, \ldots, \mathcal{M}_k\}, CK, EK)$. □

Each condition or event input connection of a module is at most connected to one incoming condition or event arc, respectively. Finally, a ${}_D$TNCES is either a self-contained Basic ${}_D$TNCEM or a self-contained Composite ${}_D$TNCEM. It has no incoming or outgoing signal interconnections, which means that the in- and output set Φ is empty. For this, it represents the highest hierarchy level of a model, which is not further composable. To compute the reachable state space of a ${}_D$TNCES, the complete set of transitions, places, and flow arcs of all modules is considered.

Definition 2.4.6 *(Complete Set of Transitions, Places, and Flow Arcs)*
Let $\mathcal{M}_S$ *be a* ${}_D$*TNCES. Then,*

- $\widetilde{T} := \bigcup_{i=1}^{k} T_{\mathcal{M}_i}$ *is the complete set of all transitions of* $\mathcal{M}_S$*,*
- $\widetilde{P} := \bigcup_{i=1}^{k} P_{\mathcal{M}_i}$ *is the complete set of all places of* $\mathcal{M}_S$*, and*
- $\widetilde{F} := \bigcup_{i=1}^{k} F_{\mathcal{M}_i}$ *is the complete set of all flow arcs of* $\mathcal{M}_S$*.*

□

All event, condition, and inhibitor connections are combined as well. These connections are characterized by their sources and sinks.

Definition 2.4.7 *(Event Connection)*
Let $\{\mathcal{M}_1, \mathcal{M}_2, \ldots, \mathcal{M}_k\}$ *be the finite and non-empty set of all* ${}_D$*TNCEM of a* ${}_D$*TNCES where* $k \in \mathbb{N}$*. Further, let* $G_E = (V_E, E_E)$ *be a directed and acyclic graph where:*

- $V_E = \widetilde{T} \cup \Big(\bigcup_{i=1}^{k} (E_{\mathcal{M}_i}^{in} \cup E_{\mathcal{M}_i}^{out}) \Big)$ *is the set of vertices, and*
- $E_E = EK \cup \left(\bigcup_{i=1}^{k} \left(EN_{\mathcal{M}_i}^{in} \cup EN_{\mathcal{M}_i}^{out} \right) \right)$ *is the set of edges.*

The set of event connections is defined by:
$\widetilde{EN} := (\bigcup_{i=1}^{k} EN_{\mathcal{M}_i}) \cup \{(t,t') : t \in T_{\mathcal{M}_i}, t' \in T_{\mathcal{M}_j}, i \neq j,\ i,j \in \{1,\ldots,k\}$ *and* $\exists$ *an irreflexive path* $<t,t'>$ *in* G_E *where* $(t,e_i) \in EN_{\mathcal{M}_i}^{out}$, $(e_j,t') \in EN_{\mathcal{M}_j}^{in}$ *and all other edges* (e_{k1}, e_{k2}) *of this path are elements of* $EK\}$, *denoted by* $t \rightsquigarrow t'$. □

Definition 2.4.8 *(Condition Connection)*
Let $\{\mathcal{M}_1, \mathcal{M}_2, \ldots, \mathcal{M}_k\}$ *be the finite and non-empty set of all* $_D$*TNCEM of a* $_D$*TNCES where* $k \in \mathbb{N}$. *Further, let* $G_{CN} = (V_{CN}, E_{CN})$ *be a directed and acyclic graph where:*

- $V_{CN} = \widetilde{P} \cup \widetilde{T} \cup \Big(\bigcup_{i=1}^{k} (C_{\mathcal{M}_i}^{in} \cup C_{\mathcal{M}_i}^{out}) \Big)$ *is the set of vertices, and*
- $E_{CN} = CK \cup \left(\bigcup_{i=1}^{k} \left(CN_{\mathcal{M}_i}^{in} \cup CN_{\mathcal{M}_i}^{out} \right) \right)$ *is the set of edges.*

The set of condition connections is defined by:
$\widetilde{CN} := (\bigcup_{i=1}^{k} CN_{\mathcal{M}_i}) \cup \{(p,t) : p \in P_{\mathcal{M}_i}, t \in T_{\mathcal{M}_j}, i \neq j,\ i,j \in \{1,\ldots,k\}$ *and* $\exists$ *an irreflexive path* $<p,t>$ *in* G_{CN} *where* $(p,cn_i) \in CN_{\mathcal{M}_i}^{out}$, $(cn_j,t) \in CN_{\mathcal{M}_j}^{in}$ *and all other edges* (cn_{k1}, cn_{k2}) *of this path are elements of* $CK\}$, *denoted by* $p \longrightarrow\!\bullet\ t$.
$W_{CN}(p \longrightarrow\!\bullet\ t) := W_{CN}(x,t) : x = p \vee x \in \bigcup_{i=1}^{k} \left(C_{\mathcal{M}_i}^{in} \right)$ *defines the weight of the condition connection.* □

Definition 2.4.9 *(Inhibitor Connection)*
Let $\{\mathcal{M}_1, \mathcal{M}_2, \ldots, \mathcal{M}_k\}$ *be the finite and non-empty set of all* $_D$*TNCEM of a* $_D$*TNCES where* $k \in \mathbb{N}$. *Further, let* $G_I = (V_I, E_I)$ *be a directed and acyclic graph where:*

- $V_I = \widetilde{P} \cup \widetilde{T} \cup \Big(\bigcup_{i=1}^{k} (C_{\mathcal{M}_i}^{in} \cup C_{\mathcal{M}_i}^{out}) \Big)$, *is the set of vertices, and*
- $E_I = CK \cup \left(\bigcup_{i=1}^{k} \left(I_{\mathcal{M}_i}^{in} \cup CN_{\mathcal{M}_i}^{out} \right) \right)$ *is the set of edges.*

The set of condition connections is defined by:
$\widetilde{I} := (\bigcup_{i=1}^{k} I_{\mathcal{M}_i}) \cup \{(p,t) : p \in P_{\mathcal{M}_i}, t \in T_{\mathcal{M}_j}, i \neq j,\ i,j \in \{1,\ldots,k\}$ *and* $\exists$ *an irreflexive path* $<p,t>$ *in* G_I *where* $(p,cn_i) \in CN_{\mathcal{M}_i}^{out}$, $(i_j,t) \in I_{\mathcal{M}_j}^{in}$ *and all other edges* (cn_{k1}, cn_{k2}) *of this path are elements of* $CK\}$, *denoted by* $p \multimap t$.
$W_I(p \multimap t) := W_I(x,t) : x = p \vee x \in \bigcup_{i=1}^{k} \left(C_{\mathcal{M}_i}^{in} \right)$ *defines the weight of the inhibitor connection.* □

Based on the signal connections, sets of predecessor and successor relations are defined.

Definition 2.4.10 *(Predecessor and Successor Relations)*
Let $\mathcal{M}_S$ be a $_D$TNCES. Then, the disjoint subsets

- $^{F}t := \{p \in \widetilde{P} : (p,t) \in \widetilde{F}\}$, $^{CN}t := \{p \in \widetilde{P} : (p,t) \in \widetilde{CN}\}$, $^{I}t := \{p \in \widetilde{P} : (p,t) \in \widetilde{I}\}$, *and* $^{EN}t := \{t' \in \widetilde{T} : (t',t) \in \widetilde{EN}\}$ *define the pre-region of a transition* $t \in \widetilde{T}$,
- $t^{F} := \{p \in \widetilde{P} : (t,p) \in \widetilde{F}\}$ *and* $t^{EN} := \{t' \in \widetilde{T} : (t,t') \in \widetilde{EN}\}$ *define the post-region of a transition* $t \in \widetilde{T}$,

The cardinal number of the set of incoming event signals of a transition t is denoted by $indeg_t^{EN} = \left|{}^{EN}t\right|$. *The cardinal number of the set of outgoing event signals of a transition t is denoted by* $outdeg_t^{EN} = \left|t^{EN}\right|$. □

A transition only having outgoing event signals is called *trigger transition.* In contrast, a transition with at least one incoming event signal is called *forced transition.* A forced transition always fire instantaneously. The state of a $_D$TNCES is given by the information of all places.

Definition 2.4.11 *(State and State Equivalence)*
Let $\mathcal{M}_S$ be a $_D$TNCES. The state of $\mathcal{M}_S$ is $z = \{(m(p), l(p)) : p \in \widetilde{P}\}$, *abbreviated to* $z = (m, l)$, *and* $z_0 = ((m_0(p_1), l_0(p_1)), \ldots, (m_0(p_m), l_0(p_m)))$, $m = |\widetilde{P}|$, *is the initial state.*
Two states z and z' are equivalent if the number of tokens and the local times of all places in z are equivalent to these ones in z'. □

The transition from one state to another is given by the semantics, which is described in the following.

2.4.2. Semantics

The semantics of $_D$TNCES is given in terms of *steps.* These are the state transitions from one discrete state to another. In an ordinary Petri net, each firing of one marking-enabled transition is one step. In a $_D$TNCES, condition- and inhibitor-enabling are considered in addition. The definition for enabled transitions is given as follows:

Definition 2.4.12 *(Marking-, Condition-, and Inhibitor-Enabled Transition)*
Let $\mathcal{M}_S$ be a $_D$TNCES and $z = (m, l)$ a state of $\mathcal{M}_S$. Then,

- $T_m = \{t \in \widetilde{T} : m(p) \geq W_F(p,t),\ ZF_R \leq l(p) \leq ZF_L,\ \forall p \in {}^{F}t \wedge K(p) \geq m(p) + W_F(t,p),\ \forall p \in t^{F}\}$ *is the set of marking-enabled transitions in state* $z = (m, l)$,

- $T_c = \{t \in \widetilde{T} : p \longrightarrow\!\!\bullet\, t \in \widetilde{CN},\ m(p) \geq W_{CN}(p \longrightarrow\!\!\bullet\, t),\ \forall p \in \widetilde{P}\}$ *is the set of condition-enabled transitions in state* $z = (m, l)$,
- $T_i = \{t \in \widetilde{T} : p \multimap t \in \widetilde{I},\ m(p) < W_I(p \multimap t),\ \forall p \in \widetilde{P}\}$, *is the set of inhibitor-enabled transitions in state* $z = (m, l)$,
- $T_{mci} = \{t \in (T_m \cap T_c \cap T_i)\}$ *is the set of marking-, condition- and inhibitor-enabled transitions,*
- $T_{trigger} = \{t \in T_{mci} : indeg_t^{EN} = 0\}$ *is the set of enabled trigger transitions, and*
- $T_{forced} = \{t \in (T_{mci} \setminus T_{trigger})\}$ *is the set enabled of forced transitions.*

□

A transition t is marking-enabled if the number of tokens on places $p \in {}^F t$ in its pre-region is $\geq$ the corresponding flow arc weights, and the capacity of places $p \in t^F$ in its post-region is $\geq$ the sum of existing tokens on $p \in t^F$ plus the corresponding flow arc weights. Furthermore, the local time $l(p)$ of places $p \in {}^F t$ must be $\geq$ the retarding and $\leq$ the limiting value of the time interval of the flow arc. Besides, it is condition-enabled if the labeling of source places $p \in {}^{CN} t$ of the associated condition connections is $\geq$ the corresponding condition connections weights. On the other hand, it is inhibitor-enabled if the labeling of source places $p \in {}^I t$ of the associated inhibitor connections is $<$ the corresponding inhibitor connections weights.
In contrast to an ordinary Petri net, an event signal synchronizes the firing of at least two transitions. This means, an enabled trigger transition $t \in T_{trigger}$ forces an enabled forced transition $t' \in T_{forced}$ to fire synchronously $\iff \exists (t, t') \in \widetilde{EN}$. The event signal is further transmitted to all connected forced transitions of t'. Therefore, all enabled forced transitions, which have an event connection to transition t, fire in one step. If one transition in this event sequence is not enabled, the sequence will be interrupted, and further transitions will be ignored.
Consequently, every step contains one enabled trigger transition and a maximal set of enabled forced transitions. Hence, a set of steps is derived for each state, which is defined as follows:

Definition 2.4.13 *(Step)*
Let $\mathcal{M}_S$ *be a* ${}_D TNCES$, $z = (m, l)$ *a state of* $\mathcal{M}_S$, $G_{Event} = (V_{Event}, E_{Event})$ *a directed and acyclic graph,* $V_{Event} = T_{mci}$ *the set of vertices, and* $E_{Event} = \{(t_1, t_2) \in \widetilde{EN} : t_1, t_2 \in T_{mci}\}$ *the set of edges.*
An executable step from state z *is the union of a singleton of one trigger transition* $t_i \in T_{trigger}$ *and a set of forced transitions, denoted by:*
$\omega(t_i) = \{t_i\} \cup \{t \in T_{forced} : \exists$ *a path* $<t_i, t>$ *in* G_{Event} *and if* $em(t) = \boxed{\wedge}$ *then* $\forall t^* \in {}^{EN} t : t^* \in \omega(t_i)\}$, $i \in \{1, ..., n\}$, $n = |T_{trigger}|$.
The set of all executable steps from state z *is denoted by* $\Omega(z) = \{\omega(t_1), \ldots, \omega(t_n)\}$.

□

If the set of forced transitions contains a transition in event mode $\boxed{\wedge}$, all enabled transitions in its event pre-region have to have the same trigger transition. Furthermore, there must only be one trigger transition in its event pre-region at most. The firing of each step leads to one successor state. Due to the dynamics, a state transition can have a step delay. The definition of both is given as follows:

Definition 2.4.14 *(Successor State, Step Delay and Conflict)*
Let $\mathcal{M}_S$ be a ${}_D$TNCES, $z = (m, l)$ a state of $\mathcal{M}_S$, and $\Omega(z)$ a set of executable steps from state z. A successor state $z'(m', l')$ of $\mathcal{M}_S$ is reached by firing a step $\omega(t_i) \in \Omega(z)$ where:

- $\forall p \in {}^{F}t,\ t \in \omega(t_i) : l'(p) = 0,\ m'(p) = m(p) - W_F(p,t),$
- $\forall p \in t^{F},\ t \in \omega(t_i) : l'(p) = 0,\ m'(p) = m(p) + W_F(t,p)$ *and,*
- $\forall p \in \left(\widetilde{P} \setminus ({}^{F}t \cup t^{F})\right),\ t \in \omega(t_i),\ m(p) > 0 : m'(p) = m(p),\ l'(p) = l(p) + 1.$

A step delay is defined by the function $\delta(z)$ where $\delta(z_0) := 0$ for the initial state z_0 of $\mathcal{M}_S$ and $\delta(z') = \begin{cases} \delta(z) + 1 & if\ \Omega(z) = \emptyset \\ 0 & otherwise \end{cases}$, where z' is the successor state of z.
A conflict is defined by the function $\mathcal{C}(\omega(t_i))$ and

$$\mathcal{C}(\omega(t_i)) = \begin{cases} true & if\ \exists p \in \widetilde{P} : m'(p) < 0 \vee m'(p) > K(p) \\ false & otherwise. \end{cases}$$ □

The successor state is derived by firing all transitions of the step $\omega(t_i)$. Doing so, the tokens on the places in the pre-regions of $t \in \omega(t_i)$ are subtracted and added to the places in the post-regions according to the flow arc weights. The local time of a place will be reset to zero if the number of tokens on it changes. If the number of tokens is at least one and does not change, the local time will be increased by one.
It is possible that $\Omega(z) = \emptyset$. In this case, there is no enabled transition and the current marking is dead. However, if there are timed flow arcs in the ${}_D$TNCES, transitions will possibly fire after a certain step delay. The delay is quantified for every state z' and specifies the number of discrete time steps before the state transition from state z to z' is executed. If $\delta(z')$ is equal to 0, the steps will fire without delay. It is reset to 0 after a non-empty step has fired.
The conflict function $\mathcal{C}(\omega(t_i))$ returns true if the number of tokens on at least one place is negative or greater than its capacity in the successor state z' of z. In this case, the elements of the power set $2^{\omega(t_i)}$ have to be considered to find the reduced maximal step(s) according to the semantics.
The complete sets of states, executable steps and step delays are combined in a dynamic graph. Such a graph corresponds to a Kripke structure [Sch04] and is given as follows:

Definition 2.4.15 *(Dynamic Graph)*
Let $\mathcal{M}_S$ be a $_D$TNCES. The dynamic graph of $\mathcal{M}_S$ is denoted by the directed graph $G_D = (Z, E_D, \mu_D)$ where:

- $Z = \{z_0, \ldots, z_m\}$, $m \in \mathbb{N}$ *is the set of states,*
- $E_D \subseteq (Z \times Z)$ *is the set of directed edges,*
- $\mu_D : E_D \rightarrow (\Omega_D \times \delta_D)$ *is a labeling function where:*
  - $\Omega_D = \bigcup_{i=0}^{m} \left(\Omega(z_i)\right)$ *is the set of all executable steps and*
  - $\delta_D = \bigcup_{i=0}^{m} \left(\delta(z_i)\right)$ *is the set of all step delays.*

An edge between the two states z and z' in G_D is a path of length 1, described by the tupel $\left(z, \omega(t_i), \delta(z'), z'\right)$ where $\omega(t_i) \in \Omega(z)$ is the executable step from state z to z' and $\delta(z')$ is the step delay of the state z'. □

An edge in a dynamic graph is given by the tupel $(z, \omega(t_i), \delta_{\Omega(z)}, z')$, which means that z' is reachable from z by firing the step $\omega(t_i)$ after a step delay of $\delta_{\Omega(z)}$. The dynamic graph is the formal basis for analyzing system models in this thesis. To do so, a formal specification of behavior is applied to it by means of a model checking software. The next section presents the specification technologies, which are deployed in this work.

2.5. Basics of Specifications

Plant requirements are documented in a technical specification. Based on it, the plant is constructed and the control software is implemented. A formal specification of behavior is further applicable to analyze the closed-loop system by means of model checking. In the following, specification techniques, which are used in this thesis, are introduced. For this, first of all temporal logics are considered. Afterwards, the theoretical background for Symbolic Timing Diagrams (STD) is provided.

2.5.1. Computation Tree Logic

Temporal logics provide a means to develop a well-formed and arbitrarily-nested specification of system's behavior. They were first proposed in [Pnu77] to describe complex temporal behavior of reactive systems. Since then, they have been successfully applied in computer science to specify and verify the behavior of discrete systems by means of model checking.

According to syntax and semantics, the set of temporal logics is grouped into *linear* and *branching temporal* logics. Both consider paths consisting of sequences of system states. The temporal information is provided considering the predecessors (the past) and successors (the future) of a certain state within this sequence. Paths start in an initial state and evolve as a directed tree. While a requirement formulated in a linear temporal logic has to hold for all paths, a branching temporal logic formula contains path quantifiers that distinguish between requirements that hold for all paths and those ones holding for at least one path. In this thesis, branching temporal temporal logics are considered. These logics are appropriate to describe most requirements regarding system behavior on the one hand and are suitable as an input language to perform model checking on the other hand. In the following, the applied temporal logics are depicted.

2.5.1.1. Syntax

The Computation Tree Logic (CTL) [BAPM83, CGP00, SR02] is a branching temporal logic. A CTL formula φ specifies the system's behavior starting in a particular state and considering paths, which evolve from this state. An *atomic expression* is a term of the form:

$$p_i = m(p_i) \; : \; p_i \in \widetilde{P}, i \in \{1, \dots, |\widetilde{P}|\}.$$

In other words, it consists of a place p_i and its labeling function $m(p_i)$. A *state predicate* is either an atomic expression, its negation, or a disjunction or conjunction of two or more atomic expressions. The simplest CTL formula consists of one state predicate. To specify temporal relations, a CTL formula is additionally composed of one path quantifier, one temporal operator and one state predicate. The path quantifier A expresses that the predicate has to be fulfilled on every path, whereas it has to be fulfilled on at least one path for the quantifier E. The temporal operator G expresses that the predicate is fulfilled in all states, for X it has to be fulfilled in the subsequent state, and for F it has to be fulfilled in a future state. Furthermore, U states that one predicate is fulfilled until a second one holds (on condition that the second predicate is fulfilled in at least one future step) and for B the predicate has to be fulfilled in one state before a second one is fulfilled in a successor state (on condition that the second predicate is fulfilled in at least one future step).
The function $P : \varphi \times Z \to \{true, false\}$ maps each state $z \in Z$ to a Boolean value. It expresses whether the CTL formula φ is fulfilled in state z or not. In Section A, the basic CTL formulas are visualized. The following definition introduces the syntax.

Definition 2.5.1 *(Computation Tree Logic Syntax [SR02])*
The set of CTL formulas is defined inductively.

Basis: *Every state predicate and the constants true and $false$ are CTL formulas.*
Step: *If φ and ψ are CTL formulas, so are the Boolean combinations $!\,\varphi$, $(\varphi \wedge \psi)$, and $(\varphi \vee \psi)$, and the temporal operators $AG\,\varphi$, $EG\,\varphi$, $AX\,\varphi$, $EX\,\varphi$, $AF\,\varphi$, $EF\,\varphi$, $A\,[\varphi\,U\,\psi]$, $E\,[\varphi\,U\,\psi]$, $A\,[\varphi\,B\,\psi]$, and $E\,[\varphi\,B\,\psi]$.*

□

2.5.1.2. Semantics

The truth value of a CTL formulas is evaluated according to a certain state of the dynamic graph, which usually is the initial state. The relation $G_D,\, z_0 \models \varphi$ expresses that the CTL formula φ is satisfied in state z_0 within the given dynamic graph G_D.

Definition 2.5.2 *(Computation Tree Logic Semantics [SR02])*
Let G_D be a dynamic graph, $z_0, z \in Z$ states of G_D, z_0 the initial state, $<z_0, z_i>$ a path in G_D between z_0 and z_i, and φ and ψ CTL formulas. Then the relation $\models$ for CTL formulas is defined inductively.

Basis: $z_0 \models \varphi \iff P(\varphi, z_0) = true$
$z_0 \models true \iff$ *always fulfilled*
$z_0 \models false \iff$ *never fulfilled*

Step: $z_0 \models\, !\varphi \iff$ *not* $z_0 \models \varphi$
$z_0 \models \varphi \wedge \psi \iff z_0 \models \varphi$ *and* $z_0 \models \psi$
$z_0 \models \varphi \vee \psi \iff z_0 \models \varphi$ *or* $z_0 \models \psi$
$z_0 \models AG\,\varphi \iff \forall <z_0, z_i>,\ \forall z : z \text{ in } <z_0, z_i> : z \models \varphi$
$z_0 \models EG\,\varphi \iff \exists <z_0, z_i>,\ \forall z : z \text{ in } <z_0, z_i> : z \models \varphi$
$z_0 \models AX\,\varphi \iff \forall <z_0, z_i>,\ \forall z : (z_0, z) \in E_D : z \models \varphi$
$z_0 \models EX\,\varphi \iff \exists <z_0, z_i>,\ \forall z : (z_0, z) \in E_D : z \models \varphi$
$z_0 \models AF\,\varphi \iff \forall <z_0, z_i>,\ \exists z : z \text{ in } <z_0, z_i> : z \models \varphi$
$z_0 \models EF\,\varphi \iff \exists <z_0, z_i>,\ \exists z : z \text{ in } <z_0, z_i> : z \models \varphi$
$z_0 \models A\,[\varphi\,U\,\psi] \iff \forall <z_0, z_i>,\ \exists z : z \text{ in } <z_0, z_i> : z \models \psi$ *and* $\forall$ *predecessor states* z' *of* $z : z'$ *in* $<z_0, z_i>$ *it holds* $z' \models \varphi$
$z_0 \models E\,[\varphi\,U\,\psi] \iff \exists <z_0, z_i>,\ \exists z : z \text{ in } <z_0, z_i> : z \models \psi$ *and* $\forall$ *predecessor states* z' *of* $z : z'$ *in* $<z_0, z_i>$ *it holds* $z' \models \varphi$
$z_0 \models A\,[\varphi\,B\,\psi] \iff \forall <z_0, z_i>,\ \exists z : z \text{ in } <z_0, z_i> : z \models \psi$ *and* $\exists$ *a predecessor state* z' *of* $z : z'$ *in* $<z_0, z_i>$ *where* $z' \models \varphi$
$z_0 \models E\,[\varphi\,B\,\psi] \iff \exists <z_0, z_i>,\ \exists z : z \text{ in } <z_0, z_i> : z \models \psi$ *and* $\exists$ *a predecessor state* z' *of* $z : z'$ *in* $<z_0, z_i>$ *where* $z' \models \varphi$

A formula φ is true in $G_D \iff$ it is true in z_0. □

2.5.1.3. Equivalences

Each CTL formula can be expressed in terms of the three operators EX, EG and EU. Consequently, the implementation costs will be reduced for model checking algorithms if those consider these operators only. The following proposition is taken from [CGP00] and extended for the operators $A\,[\,\varphi\, B\,\psi\,]$ and $E\,[\,\varphi\, B\,\psi\,]$.

Proposition 2.5.3 *(CTL Equivalences [CGP00])*
Two formulas φ_1, φ_2 *are equivalent* ($\varphi_2 \equiv \varphi_2$) *if* $z \models \varphi_1 \iff z \models \varphi_2$ *for a dynamic graph* G_D *and any state* $z \in Z$ *of* G_D.

$$
\begin{aligned}
AG\ \varphi &\equiv\ !EF\ !\varphi \\
&\equiv\ !E\ [\ true\ U\ !\varphi\ ] \\
AX\ \varphi &\equiv\ !EX\ !\varphi \\
AF\ \varphi &\equiv\ !EG\ !\varphi \\
EF\ \varphi &\equiv E\ [\ true\ U\ \varphi\ ] \\
A\ [\ \varphi\ U\ \psi\ ] &\equiv\ !E\ [\ !\psi\ U\ (\ !\varphi \wedge !\psi\ )\ ] \wedge !EG\ !\psi \\
A\ [\ \varphi\ B\ \psi\ ] &\equiv AF\ (\ \varphi \wedge AF\ \psi) \\
&\equiv\ !EG\ !(\ \varphi \wedge !EG\ !\psi\ ) \\
E\ [\ \varphi\ B\ \psi\ ] &\equiv EF\ (\ \varphi \wedge EF\ \psi\ ) \\
&\equiv E\ [\ true\ U\ (\ \varphi \wedge E\ [\ true\ U\ \psi\ ]\ )\ ]
\end{aligned}
$$

□

2.5.1.4. Rules Of Precedence

CTL formulas can be nested arbitrarily, and for this, rules of precedence are defined in order to process formulas correctly.

Definition 2.5.4 *(Rules of Precedence for CTL Formulas)*

- *! and the temporal operators AG, EG, AF, EF, AX, EX are of higher priority than*
- $\wedge$ *and* $\vee$ *are of higher priority than*
- *AU, EU, AB, EB.*

□

2.5.2. Extended Computation Tree Logic

A CTL formula specifies certain states of a system, but it is not possible to explicitly describe state transitions or steps with it. To close this gap, an extension of CTL is introduced in [SR02]. This extended Computation Tree Logic (eCTL) enlarges the expressive capability of CTL. In Section A, examples for eCTL formulas are illustrated.

2.5.2.1. Syntax

An eCTL expression enables describing state transition information, which is important particularly for specifying event-driven behavior. For this, a transition formula and the relation $\models$ are given as follows:

Definition 2.5.5 *(Transition Formula)*
The set of transition formulas is defined inductively.

Basis: *Every transition $t \in \widetilde{T}$ and the constants $true$ and $false$ are transition formulas.*

Step: *If τ and ρ are transition formulas, so are the Boolean combinations $!\,\tau$, $(\tau \wedge \rho)$, and $(\tau \vee \rho)$. The truth value of transition formulas is evaluated according to a certain edge of a dynamic graph G_D.*

□

Definition 2.5.6 *(Relation $\models$ for Transition Formulas)*
Let G_D be a dynamic graph, $(z, \omega(t_i), \delta(z'), z')$ an edge of G_D, $z, z' \in Z$ states of G_D, $\omega(t_i)$ the step from state z to z', $\delta(z')$ the step delay of z', and τ and ρ transition formulas. Then the relation $\models$ for transition formulas is defined inductively.

$$
\begin{array}{lll}
\textit{Basis:} & (z,\omega(t_i),\delta(z'),z') \models t & \Longleftrightarrow t \in \omega(t_i) \\
& (z,\omega(t_i),\delta(z'),z') \models true & \Longleftrightarrow \textit{always holds} \\
& (z,\omega(t_i),\delta(z'),z') \models false & \Longleftrightarrow \textit{never holds} \\
\textit{Step:} & (z,\omega(t_i),\delta(z'),z') \models\, !\,\tau & \Longleftrightarrow \textit{not } (z,\omega(t_i),\delta(z'),z') \models \tau \\
& (z,\omega(t_i),\delta(z'),z') \models (\tau \wedge \rho) & \Longleftrightarrow (z,\omega(t_i),\delta(z'),z') \models \tau \\
& & \quad \textit{and } (z,\omega(t_i),\delta(z'),z') \models \rho \\
& (z,\omega(t_i),\delta(z'),z') \models (\tau \vee \rho) & \Longleftrightarrow (z,\omega(t_i),\delta(z'),z') \models \tau \\
& & \quad \textit{or } (z,\omega(t_i),\delta(z'),z') \models \rho
\end{array}
$$

□

A sequence of states, whose state transitions fulfill a particular transition formula τ is defined as follows:

Definition 2.5.7 *(τ-Sequence)*
Let G_D be a dynamic graph and Z the set of states of G_D. A τ-sequence for a transition formula τ is a path $<z_0, z_k>$ in G_D where $\forall\, z_j \in Z,\ 0 \leq j < k\ \exists$ an edge $(z_j, \omega(t_i), \delta(z_{j+1}), z_{j+1})$ in G_D and $(z_j, \omega(t_i), \delta(z_{j+1}), z_{j+1}) \models \tau$, denoted by $<z_0, z_k>^\tau$.

□

Considering state transitions can limit the set of paths that have to be analyzed because only τ-sequences are taken into account. Finally, the syntax of eCTL is defined below.

Definition 2.5.8 *(Extended Computation Tree Logic Syntax)*
The set of eCTL formulas is defined inductively.

Basis: *Every predicate or atomic state proposition P and the constants true and $false$ are eCTL formulas.*

Step: *If φ and ψ are eCTL formulas, so are the Boolean combinations $!\,\varphi$, $(\varphi \wedge \psi)$, and $(\varphi \vee \psi)$, and the temporal operators $A\,\tau\,G\,\varphi$, $E\,\tau\,G\,\varphi$, $A\,\tau\,X\,\varphi$, $E\,\tau\,X\,\varphi$, $A\,\tau\,F\,\varphi$, $E\,\tau\,F\,\varphi$, $A\,\tau\,[\varphi\,U\,\psi]$, $E\,\tau\,[\varphi\,U\,\psi]$, $A\,\tau\,[\varphi\,B\,\psi]$, and $E\,\tau\,[\varphi\,B\,\psi]$, for transition formulas τ.*

□

2.5.2.2. Semantics

The semantics builds up from Definition 2.5.2. Instead of a path, a τ-sequence is taken into account.

Definition 2.5.9 *(Extended Computation Tree Logic Semantics)*
Let G_D be a dynamic graph, $z_0, z \in Z$ states of G_D, z_0 the initial state, τ a transition formula, $<z_0, z_i>^\tau$ a τ-sequence in G_D between z_0 and z_i, and φ and ψ eCTL formulas. Then the relation $\models$ is defined inductively.

$$z_0 \models A\,\tau\,G\,\varphi \iff \forall <z_0, z_i>^\tau,\ \forall z : z \text{ in } <z_0, z_i>^\tau : z \models \varphi$$
$$z_0 \models E\,\tau\,G\,\varphi \iff \exists <z_0, z_i>^\tau,\ \forall z : z \text{ in } <z_0, z_i>^\tau : z \models \varphi$$
$$z_0 \models A\,\tau\,X\,\varphi \iff \forall <z_0, z_i>^\tau,\ \forall z : (z_0, z) \in E_D : z \models \varphi$$
$$z_0 \models E\,\tau\,X\,\varphi \iff \exists <z_0, z_i>^\tau,\ \forall z : (z_0, z) \in E_D : z \models \varphi$$
$$z_0 \models A\,\tau\,F\,\varphi \iff \forall <z_0, z_i>^\tau,\ \exists z : z \text{ in } <z_0, z_i> : z \models \varphi$$
$$z_0 \models E\,\tau\,F\,\varphi \iff \exists <z_0, z_i>^\tau,\ \exists z : z \text{ in } <z_0, z_i> : z \models \varphi$$
$$z_0 \models A\,\tau\,[\varphi\,U\,\psi] \iff \forall <z_0, z_i>^\tau,\ \exists z : z \text{ in } <z_0, z_i>^\tau : z \models \psi \text{ and } \forall \text{ predecessor states } z' \text{ of } z : z' \text{ in } <z_0, z_i>^\tau \text{ it holds } z' \models \varphi$$
$$z_0 \models E\,\tau\,[\varphi\,U\,\psi] \iff \exists <z_0, z_i>^\tau,\ \exists z : z \text{ in } <z_0, z_i>^\tau : z \models \psi \text{ and } \forall \text{ predecessor}$$

$$\text{states } z' \text{ of } z : z' \text{ in } <z_0, z_i>^\tau \text{ it holds } z' \models \varphi$$

$$z_0 \models A\tau\,[\varphi\, B\, \psi] \iff \forall <z_0, z_i>^\tau,\ \exists z : z \text{ in } <z_0, z_i>^\tau : z \models \psi \text{ and } \exists \text{ a predecessor}$$
$$\text{state } z' \text{ of } z : z' \text{ in } <z_0, z_i>^\tau \text{ where } z' \models \varphi$$

$$z_0 \models E\tau\,[\varphi\, B\, \psi] \iff \exists <z_0, z_i>^\tau,\ \exists z : z \text{ in } <z_0, z_i>^\tau : z \models \psi \text{ and } \exists \text{ a predecessor}$$
$$\text{state } z' \text{ of } z : z' \text{ in } <z_0, z_i>^\tau \text{ where } z' \models \varphi$$

A formula φ is true in G_D $\iff$ it is true in z_0. □

Proposition 2.5.10 *The temporal operators of CTL, that is $AG\,\varphi$, $EG\,\varphi$, $AX\,\varphi$, $EX\,\varphi$, $AF\,\varphi$, $EF\,\varphi$, $A\,[\varphi\, U\, \psi]$, $E\,[\varphi\, U\, \psi]$, $A\,[\varphi\, B\, \psi]$, and $E\,[\varphi\, B\, \psi]$ can be derived by setting $\tau \equiv true$.*

□

Corollary 2.5.11 *CTL is a subset of eCTL.* □

2.5.3. Timed Computation Tree Logic

The already presented logics lack in describing quantitative temporal correlations. Because of this, a further extension of CTL is presented in [SR02]. The authors propose the Timed Computation Tree Logic (TCTL), which allows expressing time constraints in terms of discrete intervals.

2.5.3.1. Syntax

To quantify the accumulated step delays, a path delay is defined as follows:

Definition 2.5.12 *(Path Delay)*
Let G_D be a dynamic graph, $z_k \in Z$ a state of G_D, $z_0 \in Z$ the initial state of G_D, and $\delta(z_k) \in \delta_D$ the step delay of z_k. The path delay of a path $<z_0, z_k>$ in G_D is defined by the function $D(<z_0, z_k>)$ where:

$$D(<z_0, z_k>) = \begin{cases} 0 & if\ z_0 = z_k \\ \sum\limits_{i=0}^{k} \delta(z_i) + |<z_0, z_k>| & otherwise. \end{cases}$$ □

In other words, $D(<z_0, z_k>)$ returns the number of discrete time units after which the state z_k on the path $<z_0, z_k>$ starting in state z_0 is reached. For this, all delays of the steps on the path and the path length are accumulated. Doing so, the path delay in an untimed net corresponds to the number of state transitions.
A formula of TCTL is obtained from a CTL formula by attaching discrete intervals.

Such intervals are given by $[l, h]$ where $0 \leq l \leq D[<z_i>, z] \leq h \leq \infty$ where $l, h \in \mathbb{N}_0$. For example, a TCTL formula $EX_{[l,h]}\,\varphi$ will be satisfied by a state z if this state has a successor z' satisfying the formula φ and if the state transition from state z to z' takes at least l and at most h discrete time units.
The syntax of TCTL is analog to CTL (see Definition 2.5.1), except the attachment of intervals $[l, h]$ to the modalities X, F and U.

2.5.3.2. Semantics

The semantics builds up from Definition 2.5.2 and is extended for the temporal operators X and F in the following.

Definition 2.5.13 *(Timed Computation Tree Logic Semantics)*
Let G_D be a dynamic graph, $z_0, z \in Z$ states of G_D, z_0 the initial state, $<z_0, z_i>$ a path in G_D between z_0 and z_i, $D(<z_0, z_i>)$ the path delay of z_i, and φ and ψ TCTL formulas. Then, the relation $\models$ for TCTL formulas is defined inductively.

$$z_0 \models AX_{[l,h]}\,\varphi \Longleftrightarrow \forall <z_0, z_i>,\ \forall z : (z_0, z) \in E_D \wedge l \leq D(<z_0, z>) \leq h : z \models \varphi$$
$$z_0 \models EX_{[l,h]}\,\varphi \Longleftrightarrow \exists <z_0, z_i>,\ \forall z : (z_0, z) \in E_D \wedge l \leq D(<z_0, z>) \leq h : z \models \varphi$$
$$z_0 \models AF_{[l,h]}\,\varphi \Longleftrightarrow \forall <z_0, z_i>,\ \exists z : z \text{ in } <z_0, z_i> \wedge l \leq D(<z_0, z>) \leq h : z \models \varphi$$
$$z_0 \models EF_{[l,h]}\,\varphi \Longleftrightarrow \exists <z_0, z_i>,\ \exists z : z \text{ in } <z_0, z_i> \wedge l \leq D(<z_0, z>) \leq h : z \models \varphi$$

A formula φ is true in $G_D \Longleftrightarrow$ it is true in z_0. □

Proposition 2.5.14 *The temporal operators of CTL, that is $AX\,\varphi$, $EX\,\varphi$, $AF\,\varphi$, $EF\,\varphi$ can be derived by setting $l = 0$ and $h = \infty$.* □

Corollary 2.5.15 *CTL is a subset of TCTL.* □

2.5.4. Symbolic Timing Diagrams

Symbolic Timing Diagrams were first presented in [SD93]. They are derived from the notion of classical timing diagrams as used by hardware designers. In contrast to these diagrams, which are often used informally or with ad-hoc semantics in mind, STD have a precise semantics, which is defined by translation into a characterizing temporal logic formula [KS95]. For this, the semantics is given by unwinding the single diagrams into partially ordered symbolic automata and for this type of automata, a translation to linear temporal logic formulas is defined. Details according to the definitions of STDs and their translation to temporal logic formulas are provided in [Sch01, Wit05].

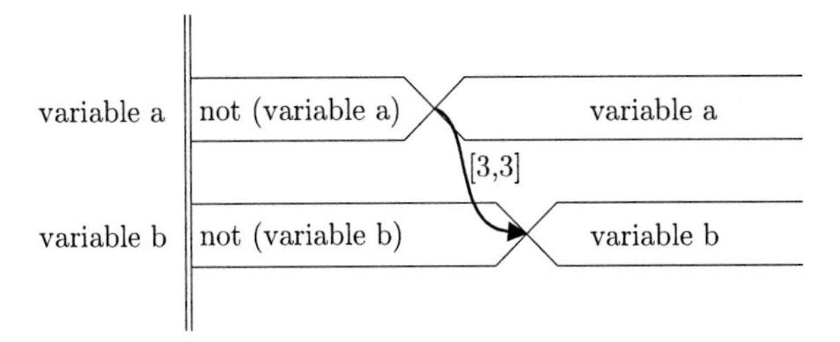

Figure 2.3.: Example for a Symbolic Timing Diagram [Wit05].

Figure 2.3 shows an example for an STD. The diagram consists of two waveforms, one for *variable a* and one for *variable b*. The two vertical parallel lines specify the initial activation of a diagram, which means that the diagram will be violated if *not (variable a)* and *not (variable b)* do not hold initially. Otherwise, the diagram will be fulfilled, if either *not (variable a)* and *not (variable b)* holds forever or if *variable b* becomes true exactly 3 steps after *variable a* has become true. Any other sequence would violate the diagram. Besides the initial activation mode, a diagram can be in iterative activation mode as well. In this case, the diagram is activated repeatedly during a system run, whenever the activation condition evaluates to true and no other instance of the same diagram is already active.

An STD is interpreted from left to right and makes an assertion about the qualitative temporal progress of a specified sequence. The state transition of a waveform, which is the change of its variable value, is called *symbolic event*. For this, a symbolic waveform specifies a required or expected sequence of symbolic events, each one defining a particular change of a value of one or more variables to which the diagram refers. Since STDs make no assertion about quantitative time, the temporal relationship between symbolic events on different waveforms is specified by *constraints*. According to [Wit05], different constraint types are distinguished. *Symmetric constraints* are used for specification of a distance between two events. They can be used to specify simultaneity (see Figure 2.4a), to exclude simultaneity (see Figure 2.4b), or to describe a concrete distance in terms of steps (see Figure 2.4c). In contrast, *asymmetric constraints* relate a source event to a target event. There are three sub-categories of asymmetric constraints. First, *precedence constraints* (see Figure 2.5a) specify an order of events, which they refer to. In general, the order is of the form *"the target event may not be observed before (only after) the source event"*. Secondly, *leadsto constraints* (see Figure 2.5b) specify a causality relation according to the referred events. Basically, leadsto constraints require that *"if the source event is observed, the target event will have to be observed eventually after activation of the diagram"*. Because of this, leadsto constraints express rather a temporal implication than an ordering of the concerned events. Thirdly, *combined constraints* (see Figure 2.5c) are a combination

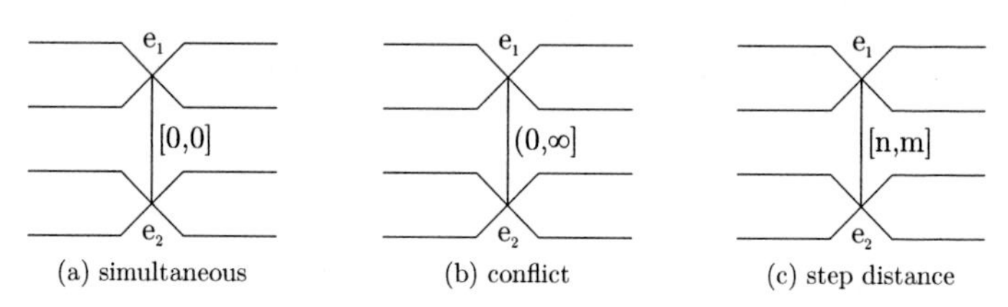

(a) simultaneous (b) conflict (c) step distance

Figure 2.4.: Symmetric constraints.

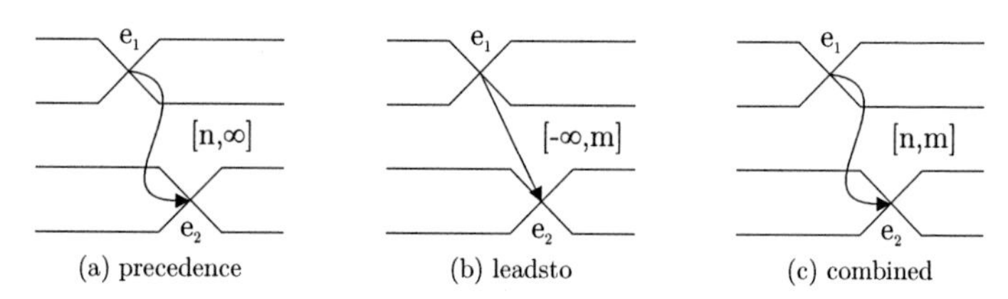

(a) precedence (b) leadsto (c) combined

Figure 2.5.: Asymmetric constraints.

of both, the ordering properties of precedence and the causality portion of the leadsto constraints.
STD constraints are either *qualitative* or *quantitative*. Qualitative constraints specify a principal temporal relation between two symbolic events, without requiring a concrete timing. This is graphically denoted by symbolic interval annotations of the constraints. In contrast, quantitative constraints are annotated with concrete time intervals. The legal intervals of constraints are depicted in Table 2.1. Furthermore, constraints are either *mandatory* or *possible*. If a mandatory constraint is violated, it will be interpreted as an error of the observed system. In contrast, possible constraints specify expectations about the temporal relation of two events. If a possible constraint is violated by a run of the system, the diagram will be canceled without returning an error. Mandatory constraints are graphically depicted by solid lines, whereas possible constraints are drawn using dashed lines.

2.6. Closed-Loop Composition

A manufacturing control system consists of a process - usually the plant - and one or more control device(s). Simulation and verification would not be complete if the controller was considered on its own. Therefore, it is natural to regard the closed-loop system of controller and plant to get a comprehensive assertion about the correctness

constraint type		valid intervals	
symmetric	simultaneous	qualitative:	$[0, 0]$
	conflict	qualitative:	$(0, \infty]$
	separation	quantitative:	$[n, m],\ n \in \mathbb{N}_0,\ m \in \mathbb{N},\ n \leq m$ $[n, \infty],\ n \in \mathbb{N}$
asymmetric	precedence	qualitative:	$[0, \infty],\ (0, \infty]$
		quantitative:	$[n, \infty],\ n \in \mathbb{N}$
	leadsto	qualitative:	$[-\infty, \infty)$
		quantitative:	$[-\infty, n],\ n \in \mathbb{N}$
	combined	qualitative:	$[0, \infty),\ (0, \infty)$
		quantitative:	$[n, m],\ n \in \mathbb{N}_0,\ m \in \mathbb{N},\ n \leq m$

Table 2.1.: Valid interval annotation of constraints [Wit05].

of the control software. As shown in Figure 2.6, controller and plant exchange information through outputs and inputs, and actuators and sensors. They do not transmit their complete state information or any common variables, but they transfer digital or analog data through their interfaces.
Workpieces represent goods like tins on a pallet, component parts or liquids. They belong to the plant, but in contrast to the plant, their properties are dynamically changed during the production process. In contrast, a plant part like a gripper may change its state from open to close, but its physical configuration remains static. Workpieces enter and leave the closed-loop system, and because of this, it is of advantage to de-

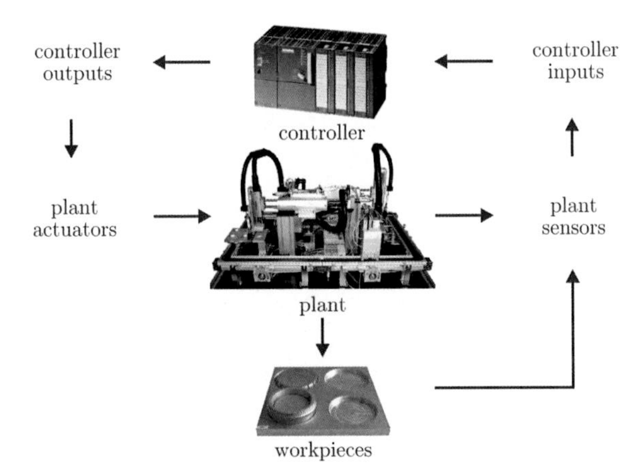

Figure 2.6.: Closed-loop system.

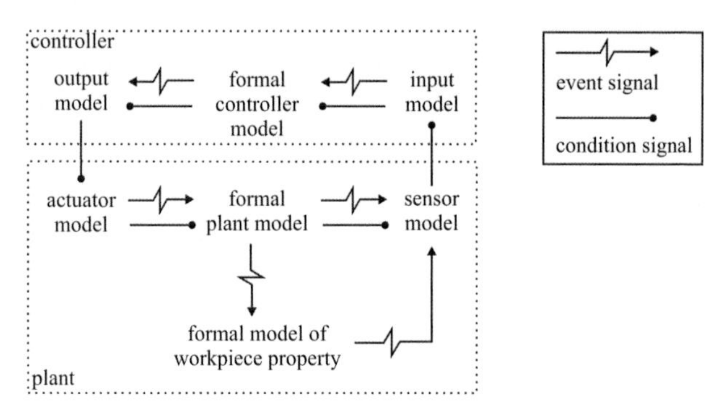

Figure 2.7.: Signal concept for the formal closed-loop system [Ger11].

scribe their properties in a separate model. They are processed by the physical plant actuators and affect plant sensors that monitor their characteristics.
Figure 2.7 shows the signal scheme of the formal closed-loop system. The two signal types, namely condition and event signals, transfer state and state transition information, respectively. The author would like to emphasize that plant and controller only are interconnected through condition signals to prevent event loops on the one hand, and to model the communication as in a real system on the other hand because sensor and actuator values represent state but not state transition information. The workpiece model is connected to the plant model through event signals. This is because the workpiece properties are directly influenced by the plant as there is a mechanical interconnection. The closed-loop system is obvious to engineers since it is natural that a controlled process needs a controller as well as a plant. However, it is daily practice that control software is written without feedback from the plant or its model. Consequently, simulating means more or less providing input information to the controller and evaluating the output information according to consistency. Doing so, the software engineers assume that they have considered every possible failure scenario based on their very own knowledge. For this, it is unnecessary to mention that open-loop testing will cause serious problems if the system complexity exceeds the human imagination for thinking about and running test cases. Consequently, it is the author's opinion that the only way to do get a comprehensive assertion about the correctness of control software is to consider both, controller and plant, in a closed loop.

2.7. Model Checking

Formal verification is a *method used to prove the correctness of a system against the specification in formal notation of its required behavior* [ISO11]. In contrast to simulation, verification considers the complete functional behavior of a system. Assertions that are made according to the fulfilling of the specification consequently hold for the whole system. This delivers a proof of correctness. Model checking [CGP00, BK08] is a verification technique and enables automatically proving properties representing the specification of a given system. All model checking algorithms can be divided into the two subsets of *explicit model checking* and *symbolic model checking*. For the sake of this work, only explicit model checking is relevant. Thereby, the specification is provided by means of temporal logic formulas.

2.7.1. General Remarks

Manufacturing control systems are a subset of the group of reactive systems [Sch04]. The environment of such a system - that is the plant - determines the points in time when an interaction is desired. For this, a reactive system has to be at least fast enough to respond to a given environment action before the next action of the environment occurs. Beyond this, it concurrently interacts with its environment and usually does not terminate. In this thesis, reactive systems are modeled with ${}_D$TNCES. To be analyzed with model checking technologies, a ${}_D$TNCES is unwound to a dynamic graph $G_D = (Z, E_D)$, which represents the explicit states and state transitions of the modeled system. Then, a formal specification of behavior in terms of a temporal logic formula f is applied to the dynamic graph. The model checking algorithm thereby determines each state $z \in Z$ and labels it with the set $label(z)$ of sub-formulas of f, which are true in z.
In practice, this procedure is automatically performed by software tools. Since there are many possibilities according to applied methods as well as input languages, a variety of software instruments supporting tool-based verification has been contributed. Because of this, the following survey regarding software tools is limited to explicit approaches and lists only a small cutout. The model checker *SPIN* [Hol97] uses an explicit approach based on automata. The system model is developed in terms of templates in *Process or Protocol Meta Language* (PROMELA). Each process template is automatically translated to a finite automaton. The specification of behavior is entered in LTL and translated to a Büchi automaton [Tri09]. SPIN is applied to verify scheduling problems of communication protocols, in particular. The authors of [CES86] present the *Extended Model Checker* (EMC), which accepts a system model provided as a Kripke structure and a specification in terms of CTL formulas. EMC was enhanced to the Symbolic Model Verifier (SMV) [McM93] in 1987. In this thesis, the algorithms of EMC are partially applied and adjusted to analyze ${}_D$TNCES,

analogously. A similar model checking approach that is used to verify communication protocols is presented in [QS82]. The software tool *CESAR* handles a system model that is given as a description program and translates it to an Interpreted Petri net. The specification is provided in a branching time temporal logic similar to CTL. The *Signal-Net System Analyzer* (SESA) [SR02] is another explicit model checker. As modeling language, it handles Signal-Net Systems (SNS), which are related to plain NCES [Kar09, Mis12]. The tool performs a reachability analysis and checks the derived graph against properties specified in CTL.
To the best of the author's knowledge, model checking of manufacturing control systems is not applied in industrial applications, today. The reasons are mainly the underlying theory and the additional costs in time and amount of work. Because of this, there is a need for integrated approaches that can be embedded into the workflow of plant engineering. To perform model checking, an explicit approach is chosen in the scope of this thesis because the algorithms are straightforward and can be implemented quickly. Beyond this, explicit model checking considers the complete state space instead of mapping it to a symbolic representation. This keeps the clarity and prevents from a loss of information. The implementation of an own model checking tool has been necessary because there was no model checker available for analyzing ${}_D$TNCES with temporal logic formulas. However, the author would like to emphasize that the applied modeling formalisms as well as the implemented algorithms can be substituted using more efficient methods. For this, this thesis is not focused on presentation of a further verification tool, but it shows that such formal methods can be embedded into the engineering practice. In the following, the applied algorithms are introduced.

2.7.2. Model Checking Algorithm

The model checking algorithms, which are depicted in this section, are inspired by those ones presented in [CGP00]. Within the framework of this thesis, the algorithms are implemented in the model checking software tool *TNCES Model Checker* (TMoC), which enables analysis of ${}_D$TNCES. Thereby, the behavior of the system is specified with CTL, eCTL and TCTL. In the following, model checking with CTL is described in detail. Nested CTL formulas are processed from the inside outwards according to the rules of precedence, which are given in Definition 2.5.4. Referring to Proposition 2.5.3, each CTL formula can be expressed in terms of the temporal operators EG, EX and EU. Because of this, it is sufficient to implement the model checking algorithms for these operators and for the Boolean terms !, $\wedge$ and $\vee$. The implication $\varphi \rightarrow \psi$ is expressed by the term $!\,\varphi \vee \psi$.
Implementing the model checking algorithm for logical terms is trivial since the values of the corresponding variables are evaluated for each state without considering the relations between the states. For formulas of the form $\omega =!\,\varphi$, each state is labeled

with ω, in which φ does not hold. In case of $\omega = \varphi \wedge \psi$, both variables have to hold in the state, and analogously, for $\omega = \varphi \vee \psi$ at least one of the variables has to hold to label the state with ω.
Evaluating temporal operators requires considering the state transitions in addition. For $\omega = EX\,\varphi$, every state, which has at least one successor that is labeled with φ, is labeled with ω (see Algorithm C.1). This procedure requires time $O\,(\,|\,E_D\,|\,)$.
To handle formulas of the form $\omega = E\,[\,\varphi\,U\,\psi\,]$, all states, which are labeled with ψ, are seeked first. Then, the algorithm goes backwards using the converse of the transition relation E_D and finds all states that can be reached by a path in which each state is labeled with φ. All such states are labeled with ω (see Algorithm C.2). This procedure requires time $O\,(\,|\,Z\,|\,+\,|\,E_D\,|\,)$.
Handling formulas of the form $\omega = EG\,\varphi$ is slightly more complicated. To narrow the state space, first of all G'_D is derived from G_D by deleting all of those states from Z, in which φ does not hold, and restricting E_D accordingly. From this follows that $G'_D = (Z', E'_D)$ where:

- $Z' = \{\, z \in Z \,|\, G_D,\, z \models \varphi \,\}$ and
- $E'_D = \{\, (\, z_1,\, \sigma,\, \delta_\sigma,\, z_2) \,\in\, E_D \,:\, z_1, z_2 \in Z' \}$.

The model checking algorithm (see Algortihm C.3) is related to the algorithm of Tarjan [AH74]. In contrast to it, the whole set of strongly-connected components is not identified as described in [CGP00], but one possible path is detected that represents a closed trajectory through the state space of Z' such that the property $EG\,\varphi$ is fulfilled. This procedure eases the complexity of the algorithm, whereas it delivers the same result as provided by [CGP00]. To evaluate the CTL formula for all states, each state $z \in Z'$ is considered as initial state. Before executing the algorithm, the set of transitions E' is stored to a temporal list T. Furthermore, the set of states fulfilling φ, that is Z', is stored to a temporal list U, and an empty list S is initialized that will contain the ordered list of states of the trajectory. T, U, and S are always reset to these values before executing the algorithm with a new initial state.
The procedure to calculate the graph is recursive. At the beginning of each call, the current state is stored to S and deleted from U. Then, the set of state transitions $\in T$ is considered. If a transition, whose source state is the current state, is found, there are two possibilities to proceed. On the one hand, if the target state is within the set of U, the transition will be deleted from T and a new instance of the procedure will be called with this state as argument. On the other hand, if the target state is within the set of S, the trajectory will be closed and the initial state will be labeled with ω. If the state is neither in U nor in S, a deadlock will be found. In this case, the state is removed from S, the resulting last state of S is moved to U and an instance of the procedure is called with this state as argument. Accordingly, the initial state will not be labeled if no closed trajectory is found. The algorithm has time complexity $O\,(\,|\,Z\,|\,+\,|\,E_D\,|\,)$.

To handle an arbitrary CTL formula ω, the labeling algorithm is applied to all sub-formulas of ω starting with the most deeply nested, and works outward to include all parts of ω. Doing so, it is guaranteed that whenever a sub-formula of ω is processed, all of its sub-formulas have already been considered. Each pass requires time $O(|Z| + |E_D|)$ and since ω has at most $|\omega|$ different sub-formulas, the entire model checking algorithm requires time $O(|\omega| \cdot (|Z| + |E_D|))$.
An example for the application of the model checking algorithm for an $_D$TNCES is presented in Section D.

2.8. Summary

This chapter introduces the basic terms and definitions of the approaches applied in this thesis. Thereby, plant and controller models are considered as well as their interconnection to the closed-loop system model. Special attention is paid to the formalisms of $_D$TNCES. Beyond this, techniques for formally specifying plant behavior are introduced. Finally, the model checking approach is depicted, and the corresponding algorithms are discussed in detail. Having established the theoretical basis for this thesis, the next chapters cover the application of theory in practice. As a starting point, the formal system modeling is considered.

3. Formal Modeling of Plant, Controller, and the Closed Loop

A model is either an abstracted representation of a real system or of another model. Its properties are influenced independently from its original. For this, test cases are applied to it without affecting the regarded system. Usually, it does not consider all details of its original but these ones that are relevant for the model purpose. Consequently, the so-called model abstraction shall reduce the complexity of analysis. This fact is crucial for analysis because it has to be ensured that the model possesses enough details to represent the real system in an adequate way. All model-based approaches rely on a well-defined and correct model. For this, all analytical results will be worthless if the model is incorrect. Because of this, a lot of efforts are made to support the process of model development [GDD09, Hir10].
In the scope of this thesis, models particularly serve for the formal verification of control software. The reasons for developing a model instead of testing properties on the real system are manifold. First of all, the real plant is usually not available, either because corresponding parts are still under construction, or the plant is in operation. In the latter case, the production cannot be interrupted because of economic reasons. In addition, tests may stress the plant components, and there exists the risk of damaging them because of software failures. The most crucial point, however, is the possibility of insufficiently-specified test cases. Especially, this concern is addressed in this thesis because the application of formal methodologies improves simulation and bridges to formal verification. In this context, the model has to be as close as possible to the real manufacturing system. For this, a modeling formalism, which is modular and hierarchical, is of advantage. In addition, it shall be time-evaluated to pay attention to the system dynamics. This time evaluation shall be discrete to ease the analysis and to keep the generated state space manageable. Furthermore, an efficient and practical signal interconnection of modules is necessary. All of these properties are fulfilled by ${}_D$TNCES, and for this reason, they are chosen as modeling formalism. However, the framework of this thesis is not limited to ${}_D$TNCES as the usage of other formalisms is supposable as well.
The more important question is how to generate models efficiently. Modeling plant and controller in addition to the actual work of plant engineering seems to be impracticable. Time-to-market pressure would prevent the integration because of additional expenses in time and money. For this, the idea of this work is to take advantage of

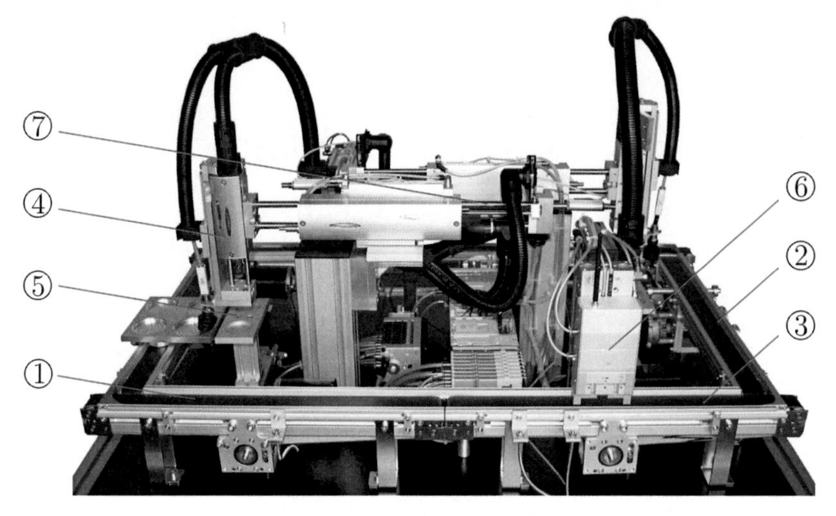

Figure 3.1.: EnAS demonstrator.

already existing data and to establish at least semi-automatic methodologies for model generation.
The biggest challenge concerning the plant is that automatic model abstraction from the real system is not possible, so far. Looking ahead, this problem will not be solved in near future. Nevertheless, a possibility is proposed to considerably improve formal plant model generation. According to controller modeling, the execution semantics as well as the actual logic transformation are to be seen as challenges. Both topics are approached in this chapter.

3.1. Demonstration Example

The framework presented in this thesis is demonstrated using the example of a manufacturing prototype. The demonstrator in Figure 3.1 is part of the *Energy-Autarkic Actuator and Sensor System* (EnAS) project[1], which deals with energy self-sufficient actuators and sensors.
The testbed consists of two identical plant modules, which are rotated by 180° to each other. The processing stations of the testbed are the *Jack Station* ④, the *Slide Station* ⑤, the *Gripper Station* ⑥, and the *Store Station* ①. To emphasize the effort of

[1]EnAS project website. URL, http://www.energieautark.com, November 2013.

applying the EnAS approach, only the right plant module is equipped with a wireless communication system and self-sufficient actuators. The left one is wired conventionally. The conveyors (①,②,③) of both modules form a circuit and cyclically transport pallets to each processing station in the clockwise direction. Each conveyor has a separate drive so that they can be moved independently from each other. For demonstration purposes, the production process is cyclic. That means tins are continuously loaded and unloaded during the production process.

Figure 3.2.: Pallet.

The pallet in Figure 3.2 can be empty, or transport one or two tins. Because of this, there are two possible positions in front of each processing station. The tins can either:

- be open,
- contain a workpiece and a loosely-closed lid, or
- contain a workpiece and a fixed lid.

The first conveyor ① transports a pallet to the first photo sensor, that is the wait position. After the sensor is activated, the controller switches off the drive and the pallet is stopped. If the drive is started again, the pallet will move on, and the sensor will be disabled. The second conveyor ② transports the pallet by means of two photo sensors to the correct positions in front of the Jack Station ④. If the first photo sensor is activated, the pallet will be in the first loading position. Respectively, the pallet will be in the second loading position if both photo sensors are activated. The third conveyor ③ transports the pallets to the Gripper Station ⑥. As for the Jack Station, the pallet will be in the first position if the first sensor is activated, and in the second if both are activated. After having finished processing, the drive of conveyor ③ is switched on. Consequently, the two photo sensors are deactivated after a certain time. The pallet will be transported to the wait position of the following plant module if the drive of its first conveyor ① is switched on.
The Jack Station puts workpieces from the Slide Station ⑤ to the tins on the pallet or vice versa. Furthermore, it can open a tin and deposit its lid onto the pallet or back onto the tin again. To do this, the vacuum gripper of the Jack Station moves vertically to the upper and lower position as well as horizontally to the three different positions retracted, middle position and extended. The middle position will be reached if the Jack Station extends an additional Jack Stick ⑦. By means of vacuum, the lid or the workpiece is ingested and lifted while the vacuum gripper moves up. Through the combination of these possibilities, the tins can be loaded and unloaded, respectively.
The Gripper Station is displayed in Figure 3.3 in more detail. Since the demonstrator's complexity would be too large for presentation in this thesis, the application of the following approaches is shown just for this substation. Nevertheless, the case studies

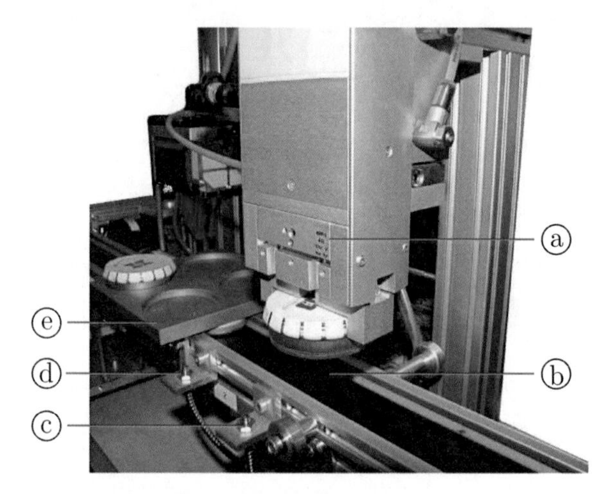

Figure 3.3.: Gripper Station.

are performed considering the whole plant. The picture shows a gripper ⓐ, a conveyor ⓑ, a pallet ⓔ, and two binary photo sensors for the handling positions. Thereby, in position 1 the sensor ⓒ is activated and sensor ⓓ is deactivated. Accordingly, sensors ⓒ and ⓓ are both activated in handling position 2. The purpose of the Gripper Station is to tightly close the lid of a tin, which was previously loaded by the Jack Station. Furthermore, a tin can be lifted and put down to another pallet. The gripper is moved up and down pneumatically and is equipped with two sensors for the upper and lower position. In Figure 3.3, the pallet has left position 2 and is moving on to the next processing station. The gripper has gripped a tin and waits for the next pallet approaching to lay down the tin on a free pallet slot.

3.2. Formal Plant Modeling

The quality of plant models is crucial for the success of closed-loop simulation and verification. Of course, the applied model has to be as abstract as possible and as detailed as necessary to be usable for analysis. However, meeting these demands does not necessarily guaranty a good model quality. Beyond this, additional work expenses has to be justified since model development requires expertise as already discussed in Section 1.1. Focusing on formal plant modeling, this section presents an approach to develop hierarchical and component-based plant models for the closed-loop analysis with $_D$TNCES.

3.2.1. Modeling of Components

Each manufacturing system consists of several mechatronic components such as conveyors, cylinders, heaters, tanks, storages, etc. as well as of digital and analog sensors and actuators. In principle, each basic component is used over and over again in the same or in other plants, and so, a systematic engineering methodology that pays attention to this point is appropriate. Each object or module shall encapsulate the discrete and uncontrolled behavior of the represented mechatronic component. Doing so, every physically-possible state of the plant component has to be modeled for each possible actuator signal configuration.
In the following, the bottom-up modeling of the Gripper Station, which is shown in Figure 3.3, is described in detail. The intention is to give an idea about the application of ${}_D$TNCES on the one hand, and to show the benefits of component-based and hierarchical modeling on the other hand. Since all components are reusable, they are combined in a library of frequently-used ${}_D$TNCE Modules [Ger11]. That means once generated, they are reusable for modeling other plants as well.
To model a plant component, first of all its physical structure has to be examined. The gripper in Figure 3.3 is composed of two pneumatic cylinders to move up and down, and to get opened and closed. The corresponding valves are controlled through two actuators for up and down movement, and one actuator for shutting the gripper. As a feedback for the controller, the gripper further possesses two end position sensors for the upper and the lower position. With this information, the formal plant module is developed.
Figure 3.4 displays the ${}_D$TNCE Module of a cylinder containing three places and four transitions. The flow arcs have an arc weight of one. Since initially there is only one token, the modeled cylinder can be in one of the three states *Extended*, *OFF-Move*, and *Retracted*. For the two flow arcs from *OFF-Move* to *t1* and *OFF-Move* to *t4*, a time interval is given to delay the firing of the corresponding transitions. Doing so, attention is paid to the dynamics of the cylinder. In the real plant, moving it from one end position to the other does not happen immediately. There are intermediate states, which could be modeled by inserting additional places. However, the actual

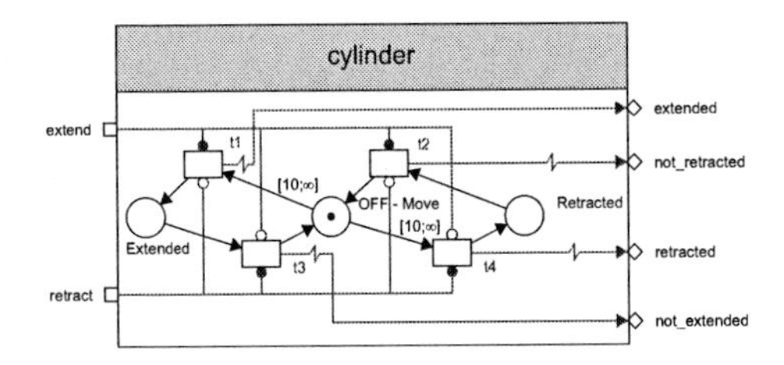

Figure 3.4.: Cylinder ${}_D$TNCEM.

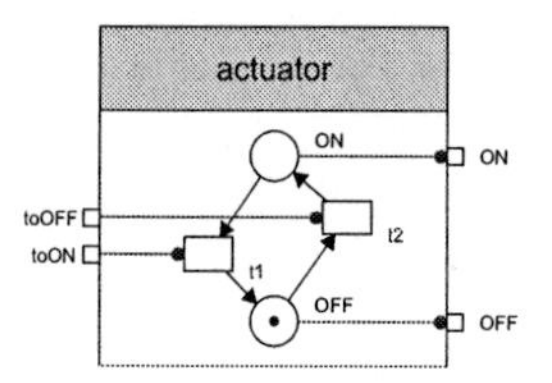

Figure 3.5.: Actuator $_D$TNCEM.

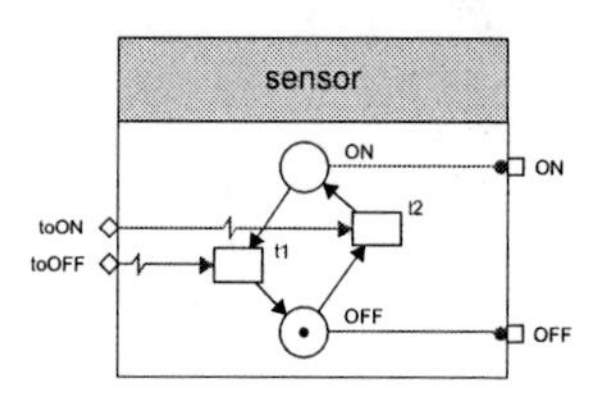

Figure 3.6.: Sensor $_D$TNCEM.

information for the controller would not be increased, but the computational complexity would rise. For this, the cylinder model is kept very abstract and dynamics are considered in terms of discrete time intervals instead of adding additional places. Of course, the intervals have to be technically meaningful since time evaluation adds computational complexity as well. Nevertheless, a case study, which is performed in this thesis, shows that temporal information in fact supports efficient modeling and analysis. Referring to the formal plant model, the author would like to emphasize that only plant components feature time intervals because these are the moving parts of a manufacturing system. The cylinder module has got the two condition signal inputs *extend* and *retract*, which are connected to the actuators modules. For this, the four transitions *t1* to *t4* fire according to the signal states of the input condition signals. The state transition information of each transition in the cylinder module is passed by event signals.
Figures 3.5 and 3.6 show the $_D$TNCE Modules of a binary actuator and a binary sensor. Both can be in state *ON* or *OFF*. According to the signal interconnection of plant and controller, which is explained in Section 2.6, an actuator gets its input information from the controller (model) through condition signals. Its state information is transmitted to the plant module, which is the cylinder. This is done by condition signals again to pay attention to the dynamics. In the real plant, the actuator is a relay, which opens and closes a valve. Hence, the air pressure for the cylinder movement is supplied. All these actions require a certain time. For this, state transitions do not occur synchronously but in terms of state sequences. In contrast, there is a physical connection between the cylinder and the sensors. This means if the cylinder reaches or leaves an end position, the sensors will change their states meanwhile. Because of this, the sensor is connected through event signals to the plant module. According to Section 2.6, the state information of a sensor module is transmitted through condition signals to the controller (model), which closes the loop.
The hierarchical composition of cylinders as well as of actuators and sensors to the *gripper* module is illustrated in Figure 3.7. It consists of the three actuator modules *move_down*, *move_up*, *shut*; the two cylinders modules *up_down*, *grip*; and the two sensor modules *retracted*, *extended*. The advantages of component-based and hierarchical modeling are clearly visible. The encapsulation of behavior into modules creates

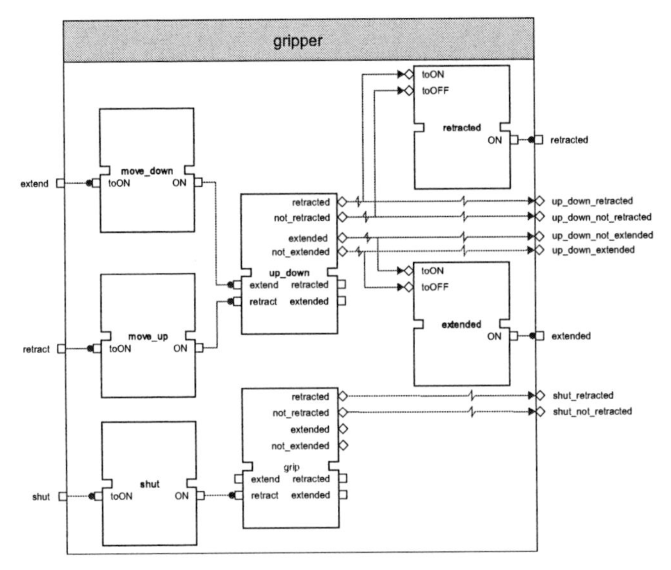

Figure 3.7.: Gripper $_D$TNCEM.

different levels of detail. For this, information is structured and does not overload the model view. In contrast, the flow of communication signals is easily comprehendible. The main benefit indeed is the implementation of a library of modules. Once generated, modeled plant components are reusable for other system models, too. In this context, a module in a library can represent either a single cylinder, or a whole functional unit like the Gripper Station.

The modules of the cylinders *up_down* and *grip* are identical. However, the *grip* module is connected with just one incoming condition arc as the represented cylinder features only one valve in the real plant. To pay attention to such open signal interconnections, the three following modes have to be implemented in a tool, which calculates the reachable states. First of all, open connections are ignored, which means that they have no impact on the connected transition. Secondly, open connections are set to false. Doing so, the connected transition will never be condition-enabled. The third possibility is to enable the user to control open connections. This feature is very practicable to test stand-alone modules.

The signal interface of the *gripper* module again corresponds to the closed-loop signal concept of Section 2.6. For this, the plant model is featured with condition in- and outputs to be interconnected with the controller model. The six event outputs are necessary to connect plant and workpiece module. Regarding the gripper model, a

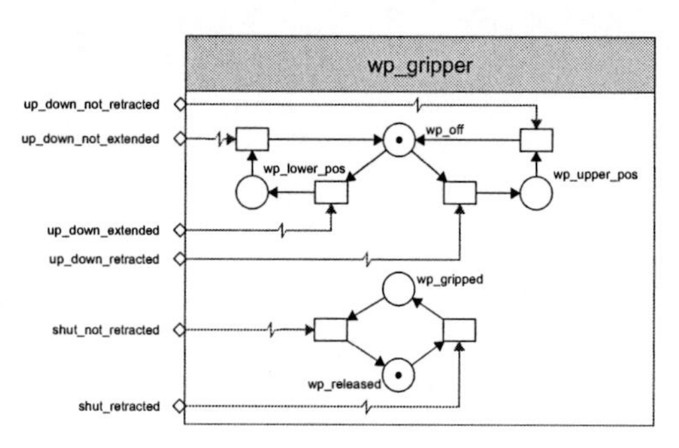

Figure 3.8.: Gripper workpiece ${}_D$TNCEM.

general issue can be seen. Whereas the position sensors provide direct feedback from the cylinder, which moves the gripper up and down, there is no sensor information about the cylinder, which shuts the gripper. Consequently, there is no feedback to the controller, whether a tin is gripped or not. This problem is only solvable with additional sensor information, and for this, it has to be attended during plant construction if necessary. Nevertheless, the actuators have a direct impact on the states of the workpieces due to their physical interconnection. To pay attention to this and especially to be able to consider it during model checking, a workpiece model is established as shown in Figure 3.8. It consists of two concurrent nets - the upper one for the vertical movement and the lower one for the gripping. For this, the workpiece can be in upper (*wp_upper_pos*), in lower (*wp_lower_pos*), or in intermediate (*wp_off*) position. Furthermore, it can be gripped (*wp_gripped*) or released (*wp_released*). The workpiece model immediately changes its states if a cylinder reaches the corresponding end position because of the mechanical connection of gripper and tin. For this, the signal interconnection is modeled with event arcs. As there is no feedback to the plant - that means the gripper does not realize if the action has been carried out successfully - there is no connection back to the plant model.
A further plant component of the Gripper Station is the conveyor. The model of the *conveyor_belt* is depicted in Figure 3.9. It receives actuator state information through the condition input *conveyor_toON*, and provides state transition information through the four event outputs. The conveyer can be in one of the four states *OFF - Move*, *Pos1*, *Pos1_2*, and *Pos2*, whereas *Pos1* and *Pos1_2* correspond to the handling positions 1 and 2. The discrete intervals model the dynamics of the conveyor belt. The event outputs are not directly connected to the sensors because like in the real plant, sensor states are not influenced by the conveyor itself but by the pallet, which moves on it. For this, more information has to be included in the model without adding artificial intelligence. Again, an appropriate approach is to additionally model the workpiece behavior. Figure 3.10 shows the corresponding implementation. According

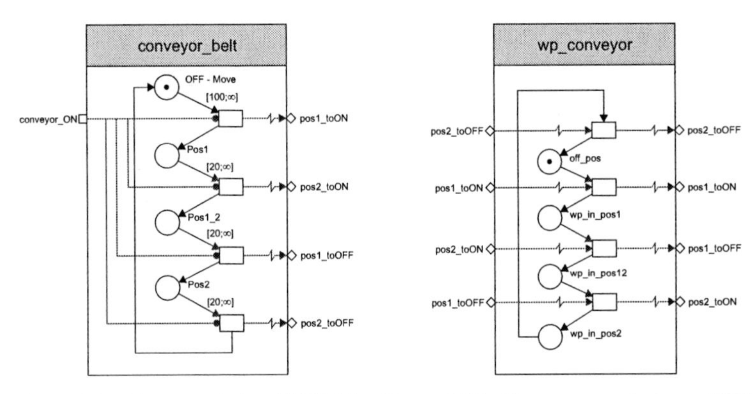

Figure 3.9.: Conveyor belt $_D$TNCEM. Figure 3.10.: Conveyor workpiece $_D$TNCEM.

to the signal concept of Section 2.6, the workpiece model is connected to the plant model only through event signals due to the direct mechanical interconnection. The four transitions are triggered by incoming event signals, which are fired by the transitions of the *conveyor_belt* module. For this, the workpiece model is synchronized by the plant model. Each state transition of the *wp_conveyor* module triggers an event signal, which is transferred to a sensor through event outputs. Doing so, the sensor state transitions are forced. Initially, the pallet is in an undefined position (*off_pos*). If the conveyor starts, the pallet reaches handling position 1 (*wp_in_pos1*). Next, the

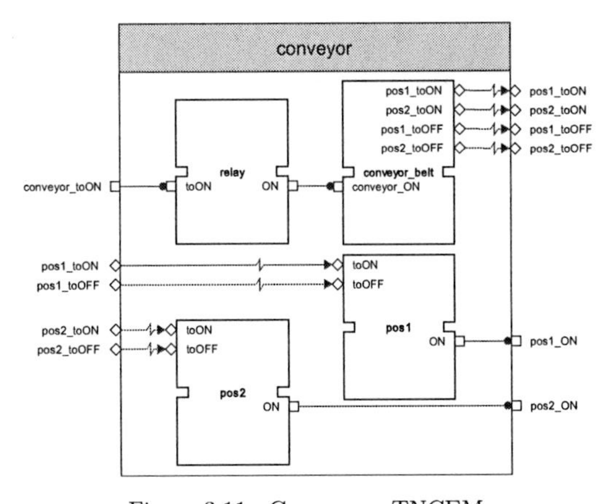

Figure 3.11.: Conveyor $_D$TNCEM.

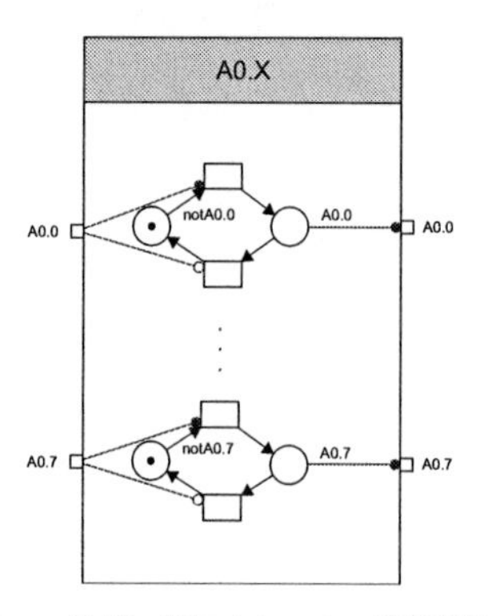

Figure 3.12.: Plant input $_{D}$TNCEM.

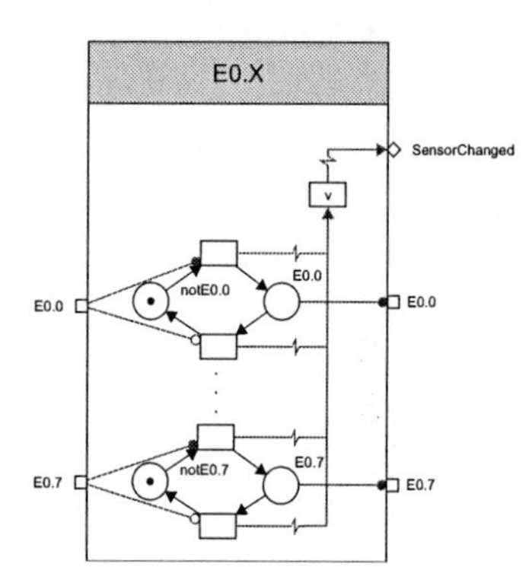

Figure 3.13.: Plant output $_{D}$TNCEM.

pallet reaches handling position 2 (*wp_in_pos1_pos2*), and subsequently, it leaves the handling positions (*pos2_ON*). Finally, it is in *off_pos*, again.
The composed model of the conveyor is illustrated in Figure 3.11. The components are the actuator module *relay*, the plant module *conveyor_belt*, and the two position sensors *pos1* and *pos2*. It features event in- and outputs to be connected to the workpiece module. Furthermore, it has got a condition interface to the plant in- and output modules, which are shown in Figure 3.12 and Figure 3.13. The modules *A0.X* and *E0.X* are necessary for HiL Simulation and Verification. In principle, they are similar to an electrical switch case. Usually, in a plant all wires run together in such a device to ease the final cabling of plant and controller. In the formal context, the modules are applied to provide the plant input and to monitor the plant output information of the model. This means that the places *A0.0* to *A0.7* are labeled according to the controller output configuration. The other way round, the labeling information of the places *E0.0* to *E0.7* is passed to the controller inputs.
The interconnection of all plant and workpiece modules is illustrated in Figure 3.14. Although both presented workpiece models are simple, they close the gap, which results from a control strategy without direct feedback. As visible, the signals from and to the workpiece modules form a closed loop. For this, workpieces belong to the plant and do not have a direct connection to the controller. Besides the requirement to model plant components without intelligence, "outsourcing" the workpiece behavior keeps the reusability of plant modules. However, the workpiece module is process-specific, and for this, it has to be adapted for each application.
The other plant components of the demonstration example are modeled similarly. The efforts decrease the more modules are contained in the library. For this, bigger systems can be modeled by just assembling the modules hierarchically. However, going one step further, this manual work is improved as presented subsequently.

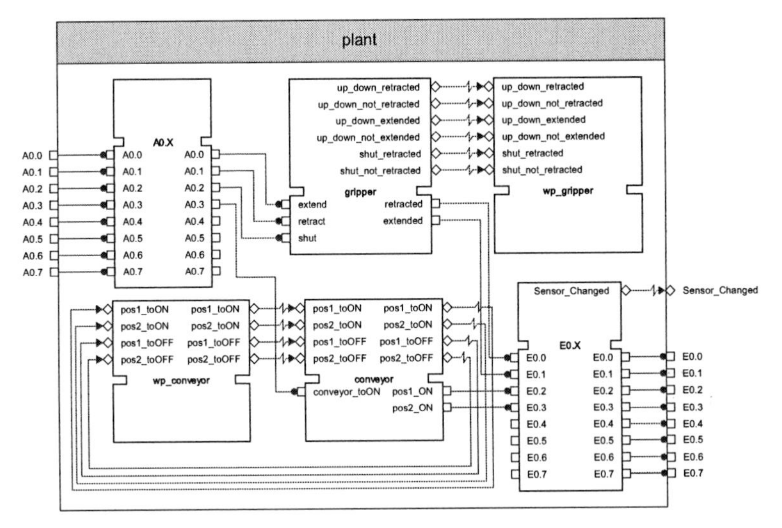

Figure 3.14.: Plant $_D$TNCEM.

3.2.2. Semi-Automatic Model Generation

Modeling is task, which is manually performed by humans. Plant models are created by abstracting from the real plant, which either is existing or is being designed. Computer-aided technologies support the model creation but they do not supersede the engineering work. The topic of this subsection is not to approach this circumstance but to point out the integrated and smart reuse of existing data.

Plant engineering is an interdisciplinary task since it is team work between engineers of different domains such as electro, process, material, or automation engineering. Usually, the work is separated to the different expert fields and joint together when it is finished. This procedure is of advantage because many development steps are performed in parallel. However, the different work packages may not match if their data basis is inconsistent. This might happen if the technical specification is ambiguous, if there are different ideas about the final plant, or if failures occur during the development process. To cope with this problem at an early project phase, it would be a great simplification to use the same basic model in all development domains or at least to ensure that different modeling environments support a neutral data exchange format so that models can be synchronized.

There are different approaches that standardize the exchange of planning data for technical plants by introducing useful meta-modeling concepts. The oil and gas industry agreed on the ISO 15926 [ISO04] standard for an overall lifecycle description.

A standard especially driven by automation is the IEC 62424 [IEC08a]. The authors of [HCG+12] compare both standards in a study and reveal specific advantages and disadvantages. Based on the IEC 62424, the data exchange format is implemented in the *Automation Markup Language* (AutomationML) [Aut12] for production automation as well as in the *Piping and Instrumentation Diagram Exchange* (PandIX) [SE12] for process automation. Summing up, the idea is common to develop only one basic plant model and to derive all other representations from it. Such a basic model can be created using a three-dimensional graphic editor, namely a CAD tool. The drawing contains all necessary information to construct the plant. However, it is static and does not contain dynamic information. The authors of [VHPY09] approach this challenge and propose the simulation model generation out of CAD data through a set of model transformations. In addition to the static model, their approach requires a *Mechatronic Object Diagram*, which is used to identify the components. The derived simulation model is provided in Simulink and translated to NCES and IEC 61499 Function Blocks for further analysis. Although the approach is very promising, the author of this thesis has the intention to limit the necessary intermediate steps. Because of this, commercial simulation tools, which support the model generation out of CAD data, are considered. There are numerous tools available for plant simulation. Nevertheless, the different environments do usually not support a standardized exchange format to enable the import and export of their models. Consequently, the automatic generation of a simulation model using CAD data is a software-specific task, which needs different parsers that support different formats of different tools. However, simulation and virtual start-up will certainly gain rising importance for the plant engineering domain. For this, it is up to the simulation software vendors to apply the guidelines and standards, which are mentioned above.
In the scope of the OMSIS project, the commercial simulation software Incontrol Enterprise Dynamics® (IED) is applied to import CAD data and to convert it to a simulation model. The software reads in a file in *CAD drawing interchange file format* (*.dxf) according to a *CAD import script* (*.cis). The CAD model of the demonstration plant is shown in Figure 3.15. The different plant parts are recognized by IED and atoms, which represent the different functional units, are derived. Doing so, a differentiation into moving and static parts is made to obtain the simulation of the plant, which is shown in Figure 3.16. The representation is simplified and shows the two-dimensional view of the demonstrator. Of course, there are no restrictions according to the design of the plant simulation model. For this, each detail is includable, and thereby, the three-dimensional simulation model would look similar to the CAD model. However, because of the complexity of the CAD models, it is sometimes suggestive to generalize them before importing them to IED. That means to abstract information like parts that are important only for the construction but not for the simulation.
In addition to the automatic CAD import, it is possible to manually import the CAD data and to define the atoms by hand. Since the simulation model is derived from the

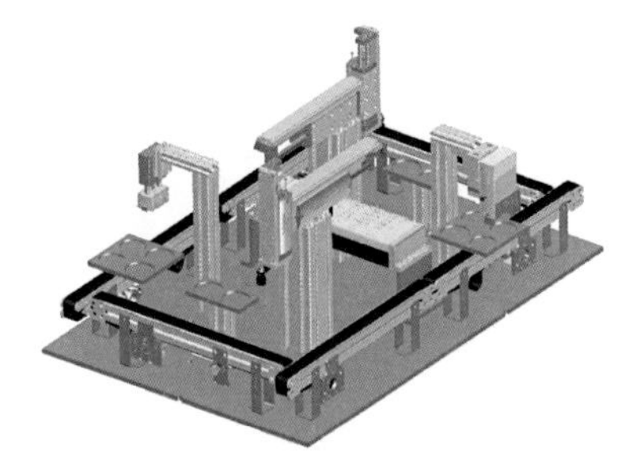

Figure 3.15.: CAD plant model.

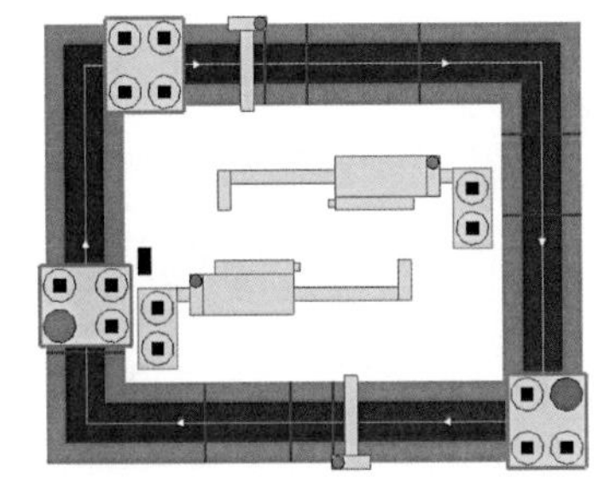

Figure 3.16.: Simulation plant model.

CAD model, it is assured that both models describe exactly the same plant. Beyond this, it is always of advantage to present a moving and colorful simulation instead of just providing the technical specification. For this, the plant construction can additionally be agreed with the customer in an early project phase.
The simulation model generation out of CAD data is a task, which is more or less supported by commercial simulation tools. For this, plant engineering companies should consider this capability when deciding on a corresponding software. However, deriving a formal model from a simulation model is not performed in manufacturing industry so far. Nevertheless, this formal model is indispensable for simulation and verification as proposed in this thesis. For this, the model again shall be derived from an already-existing representation to prevent from modeling errors. Of course, mapping a CAD model to a formal $_D$TNCES model is possible as well by applying a standard exchange format. However, if the simulation model is already available, this dynamic model will be appropriate to map it to the formal representation. For this, the formal model generation out of CAD data is not addressed in this thesis.
Due to the limited support of a standardized exchange format, which is discussed above, the model generation is very tool-specific. In principle, an own parsing tool for each simulation software has to be implemented. To systematize this task, the following restrictions for the model transformation have to be considered:

1. The most important requirement is an *open exchange format*. For this, project files must not be encrypted and shall be readable by any text editor. In ideal case, the data structure is provided in a standard format.
2. The model structure has to be *hierarchical* and *modular* so that each simulation module can be represented by a $_D$TNCEM. This request is crucial for an automatic mapping, which is based on a library of predefined modules.
3. To ease the interconnection of plant and controller (models), the formal plant model is composed of individual sensor, actuator, and plant modules. This *type-based* distinction has to be made in the simulation model as well. In addition, this requirement is also of importance to find and to define interface connections.

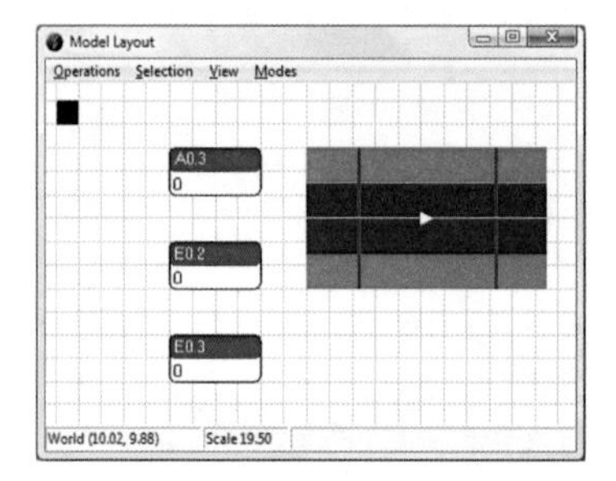

Figure 3.17.: Conveyor simulation model.

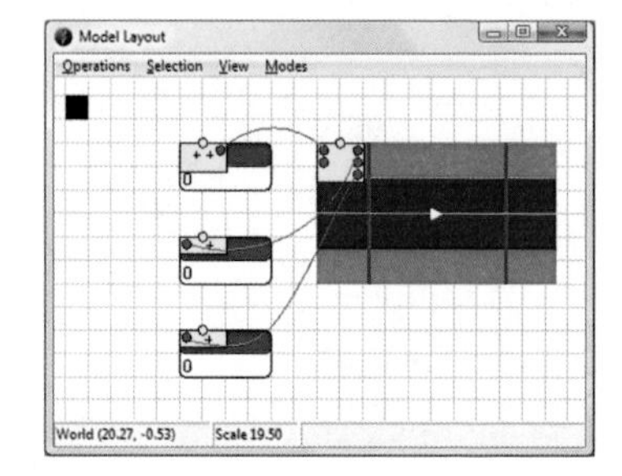

Figure 3.18.: Signal interconnections.

4. Finally, the simulation must contain the *uncontrolled behavior*, only. This means that there must be no controller logic within the simulation model.

IED meets all of these requirements. The only weak point is the specific file exchange format that makes it complicated to derive formal models automatically. Nevertheless, this problem is common to the majority of simulation software. Figure 3.17 shows the model of a conveyor in IED. To the left of it, the models of the actuator *A0.3*, and the sensors *E0.2* and *E0.3* are shown. These interface modules are further applied to interconnect plant simulation and controller hardware. Each module in IED features a communication interface as shown in Figure 3.18. The yellow boxes combine all communication channels of the module. For this, channels on the left are input channels, which receive data from other modules. Conversely, output channels on the right transmit data to other modules. The specification of each channel is provided in terms of a script. In addition, logical assignments are possible to implement the component behavior. However, this feature is not applied because according to restriction 4, the behavior is indicated only by the controller. The conveyor module possesses one input channel for the actuator and two output channels for the sensors. In addition, it has got one input and one output channel for the workpiece model. Both are connected to the conveyors of the previous and the following processing station, respectively. Doing so, the current conveyor gets the information from the conveyor of the preceding station whether a pallet is available. Accordingly, it sends this information to the conveyor of its subsequent station. However, the pallet is not modeled as a stand-alone module but as an object, which is included in every plant component. The parameters are routed, and thereby, property changes are transferred through the communication channels. Unfortunately, this makes it practically impossible to extract the workpiece information since there exists no global workpiece module. For this, such a formal model has to be created manually.
IED exports the simulation model to a *.mod file. Data is provided in plain text as claimed by restriction 1. In addition to the graphical information, the correlation of

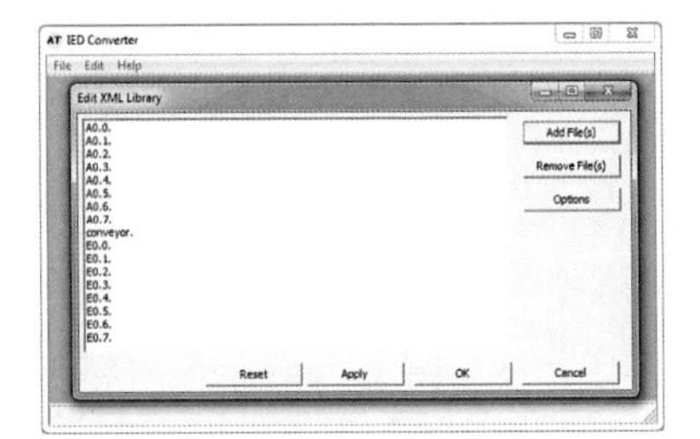

Figure 3.19.: Library in model converter.

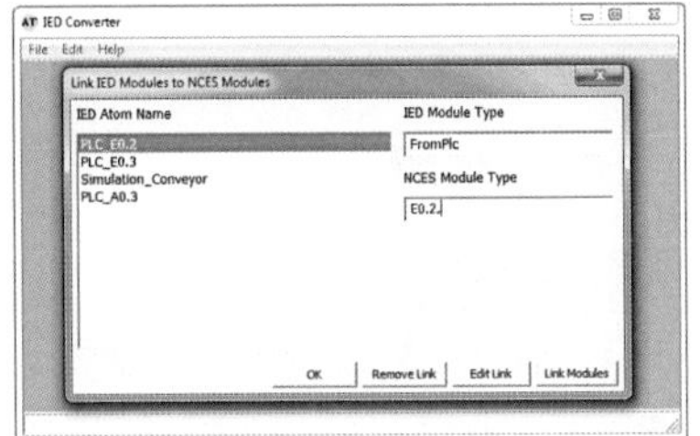

Figure 3.20.: Linker in model converter.

the different modules is provided. This information is parsed and converted to the formal $_D$TNCE plant module. To do so, the converting tool *IED Converter* is developed in the scope of this thesis. It imports the *.mod file of IED, interprets it, and maps the detected simulation plant modules to $_D$TNCEM. For this, it contains the library of predefined $_D$TNCEM, which is shown in Figure 3.19. The module types are linked in the dialog shown in Figure 3.20. The IED Converter involves the types of the modules and automatically maps similar ones. For the example in Figure 3.17, this means that the connection of one simulation input module to a $_D$TNCEM input module is done once. Afterwards, all other relations to the same module type are drawn automatically. The signal interconnection between the formal modules will be drawn automatically if it is unambiguous. That means the names of the communication channels within the simulation and the signal interfaces within the formal model have to be identical. Otherwise, the user is prompted by a dialog to define the connections manually. The connection of the plant input module to the conveyor model is shown in Figure 3.21. Finally, the model has to be exported to a $_D$TNCES. The result of the overall process for the simulation model in Figure 3.17 is shown in Figure 3.22. This model does not contain the workpiece model, so far. As stated above, a corresponding module has to be implemented manually. For this and because of the necessary dialog-based adap-

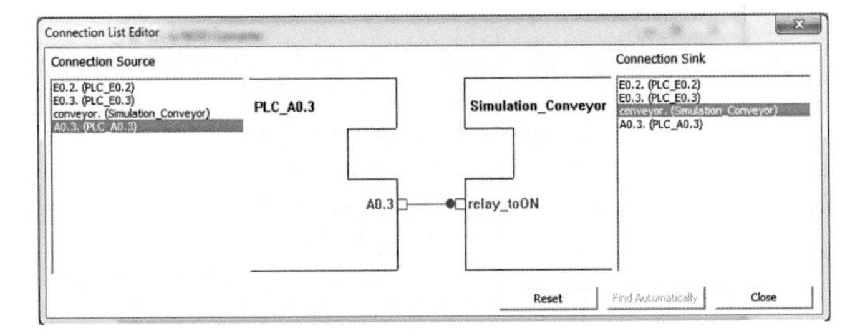

Figure 3.21.: Signal interconnections dialog.

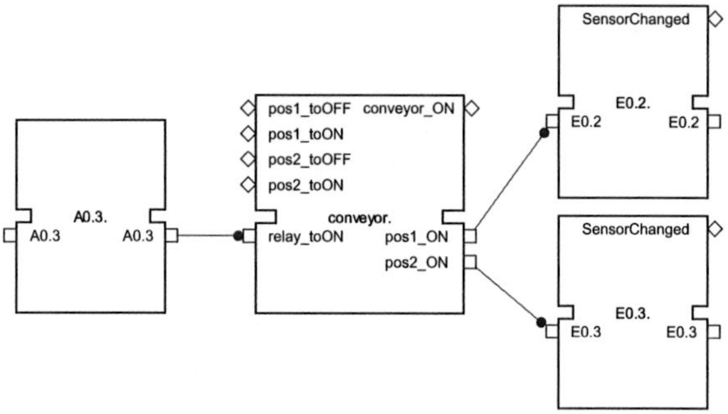

Figure 3.22.: Converted formal plant model.

tion of ambiguous information, there is still the need for user-interaction. However, this could be minimized by a standardized data format. In addition, establishing a library of formal modules is a necessary precondition for the mapping process. Consequently, this task will remain a job, which has to be done by engineers. However, if verification is more frequently applied for plant engineering in the future, hardware vendors will hopefully support this work and deliver their plant parts together with the corresponding formal modules.
Summing up this section, a possibility is provided for the further usage of the simulation model. Even though the workpiece behavior has to be modeled by an engineer, the formal model generation is accelerated. Beyond this, no information is lost since the IED Converter considers all simulation modules. To establish the SiL closed loop, of course the controller has to be modeled as well. This concern is addressed in the subsequent section.

3.3. Formal Controller Modeling

A controller model is inevitable to perform SiL Verification. However, there are further motivations to make such a modeling effort. On the one hand, the controller model in general is applicable to run test cases and to evaluate functionalities without affecting the corresponding controller. On the other hand, it is suitable for documentation and migration even after several years of plant runtime. Once developed, a formal model can be translated to control code in any language. This procedure does not depend on specific engineering environments, programming languages, or particular vendors. For this, it seems to be "ageless" as translation rules can always be defined. In the following, it is described how to create a formal controller model with $_D$TNCES. Afterwards, the generation of such a model out of already existing control code is depicted.

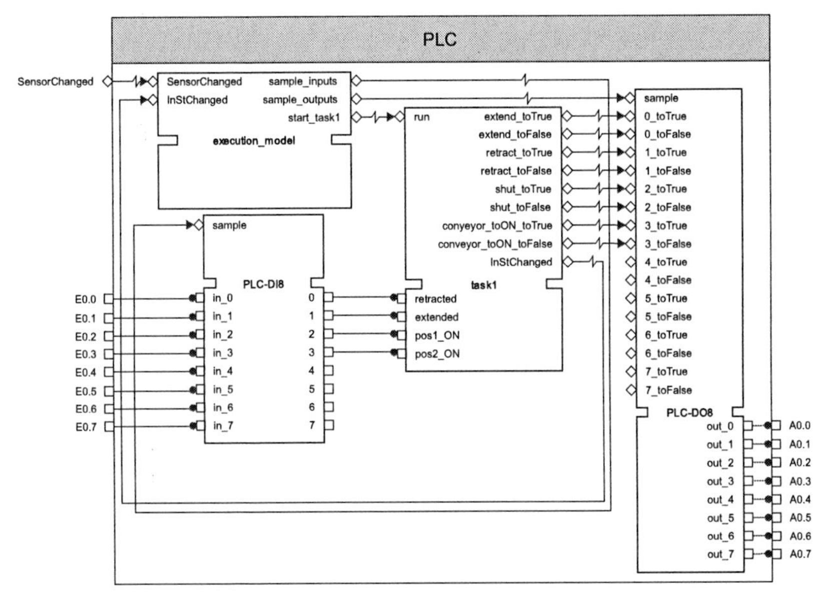

Figure 3.23.: PLC $_D$TNCEM.

3.3.1. Controller Modeling

The plant model, which is introduced in the previous section, features condition in- and outputs according to the signal concept in Figure 2.6. To close the loop, controller in- and output information is also transferred through condition signals so that both models can be interconnected. In the following, modeling of IEC 61131-3-conform controllers is considered. For more information on IEC 61499 controller modeling, the reader is referred to [PMG$^+$11, Ger11]. According to Section 2.3, a PLC has a certain execution behavior. To model preferably close to reality, this behavior must be accounted. In case of an IEC 61131-3-conform PLC, the control software cyclically runs on the device like described in Section 2.3.1. Referring to [IEC03], a PLC *configuration* consists of one or more *resources*. In turn, each resource represents one CPU and executes one or more *tasks*, which consist of one or more *programs*. On the lowest level, every program contains *function blocks* or *functions*, whereas function blocks are composed of basic functions.

As an example, the gripper and its related conveyor in Figure 3.3 are considered. Figure 3.23 depicts the $_D$TNCES of the corresponding IEC 61131 resource, which is the PLC. It consists of four modules representing the *inputs* and *outputs*, the *exe-*

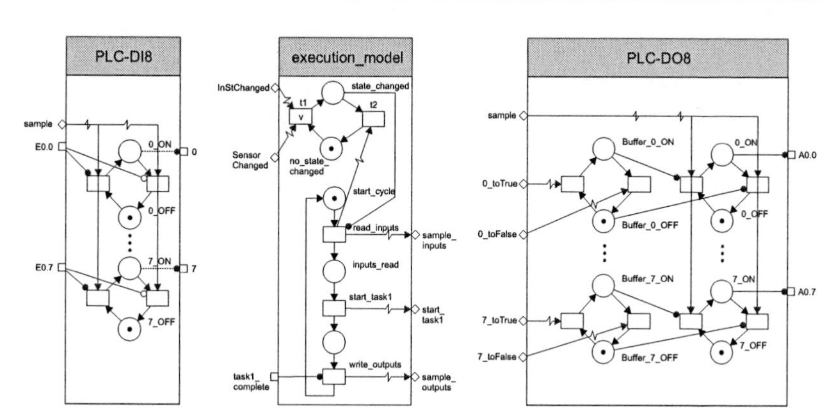

Figure 3.24.: PLC input, execution, and PLC output $_D$TNCEM.

cution model, and *the task model.* Each resource features only one execution model. It controls the cyclic processing of the control software and is depicted in the center of Figure 3.24. The module contains two subnets. The lower one models the actual execution behavior. Initially, there is a token on *start_cycle.* At the beginning of each cycle, the transition *read_inputs* is fired and polls the controller inputs by triggering the event *sample_inputs.* Afterwards, the token is on place *inputs_read.* The polled input information is stored to the process image, which is the buffer shown in the left of Figure 3.24. Its information remains stable until the inputs are sampled again. For this, data is consistent for each executed function. Subsequently, the execution model starts the task sequence by triggering the corresponding events. Output information is provided by the functions according to the controller logic. This information is stored to the process image, which is shown in the right of Figure 3.24. Each task returns a condition signal after having finished. Finally, the event *sample_outputs* is fired to write the output information from the buffer to the actual controller outputs. Then, the place *start_cycle* is labeled again.
This execution semantics corresponds to the IEC 61131 standard. The controller cyclically executes the control algorithm in its memory, which is of advantage to calculate the maximal execution time and to give an assertion according to real-time capabilities of the controller hardware. However, it is very inadequate for the formal model. After running just a few cycles, the state space is needlessly blown up. This becomes obvious considering the dynamics. A PLC executes on cycle in the range of milliseconds. In contrast, the manufacturing plant usually provides new information in the range of seconds. Consequently, hundreds or thousands of cycles are executed before a sensor state transition is actually recognized. To approach this characteristic, the *execution_model* is extended by the subnet in the upper part of the model in Figure 3.24.

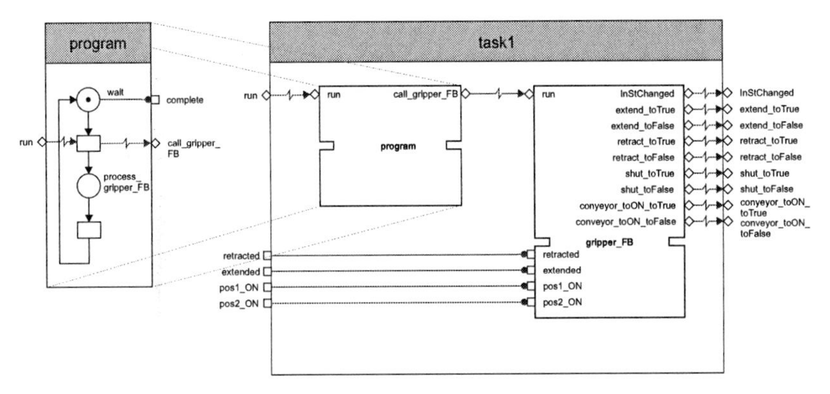

Figure 3.25.: Task $_D$TNCEM.

Initially, this subnet is in state *no_state_changed*. If a function requires an update, which means if for example a timer is elapsed and requests to change the output states, the event *InStChanged* will be triggered. The event *SensorChanged* is triggered by a sensor module within the plant model, which is shown in Figure 3.12. For this, the transition *t1* will fire and the place *state_changed* will be labeled *only if* an update is requested. In this case, the condition for firing the transition *read_inputs* is fulfilled, and the execution cycle is started. Concurrently, the transition *t2* fires and the place *no_state_changed* is labeled again. The author would like to emphasize that this simplification is not present in the real controller. However, it is necessary to prevent the formal closed-loop model from generating dispensable states by iteratively running the cycles even if no information is provided. As visible in the *execution_model*, transition *t1* is an event sink. For this, an event loop is avoided a priori. In any case, each state transition is recognized by the approach so that no information is lost.
The task module is depicted in Figure 3.25. It consists of one *program* module, which calls *function block* modules containing basic *functions*. For this, the *program* module organizes the sequential execution by firing *call_FB* events. It does not have an interface to the controller in- and outputs. In contrast, the *gripper_FB* module contains the actual controller logic and provides output information by processing input signals and internal state information. To do so, the function block modules are directly interconnected with the process image. The controller logic module for the gripper and the conveyor is given in Figure 3.26. It features an event input *run*, which triggers the execution, and condition inputs for the input variable values. On the other side, it provides the event output *InStChanged* and the event outputs for the output variable values. The transitions *t1* to *t8* just merge the events since each module output must only be connected with one arc per definition. The sequence

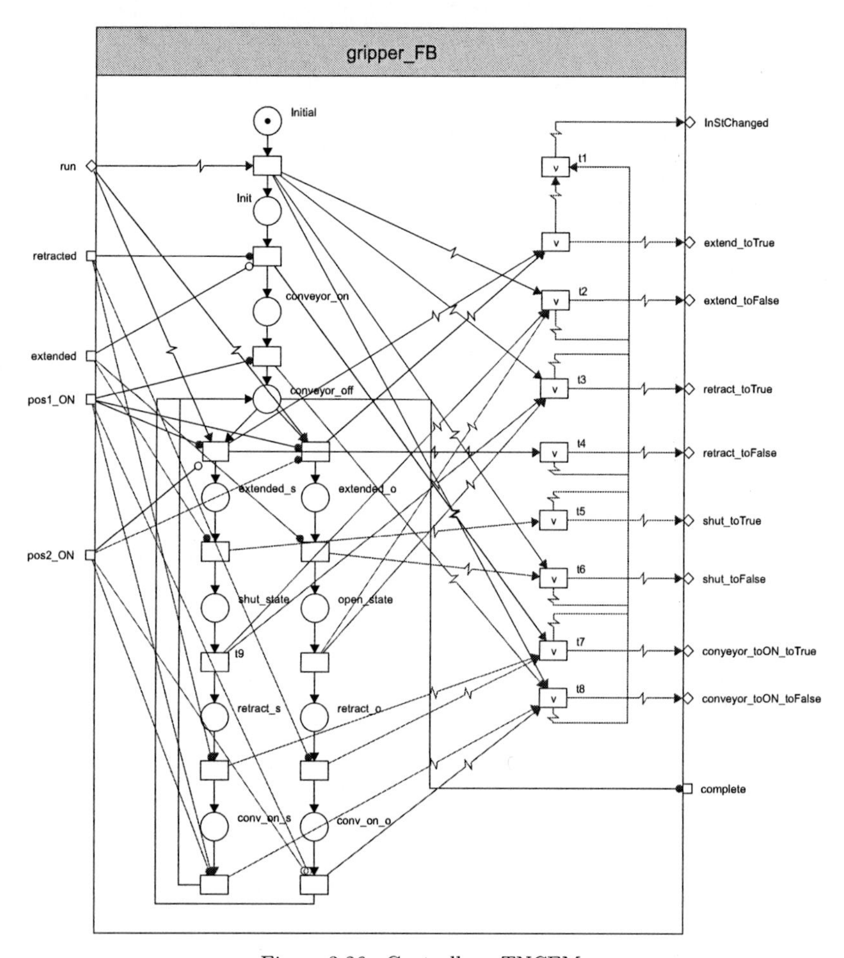

Figure 3.26.: Controller $_D$TNCEM.

starts in *Initial* state. After the *run* event fires, the gripper is opened (*shut_toFalse*) and raised (*retract_toTrue* and *extend_toFalse*). Then, the conveyor is switched on until a pallet reaches handling position 1 (token on place *conveyor_on*). Next, the conveyor is switched off and the function block is in *conveyor_off* state. After initialization, the actual production sequence will start if the *run* event fires. According

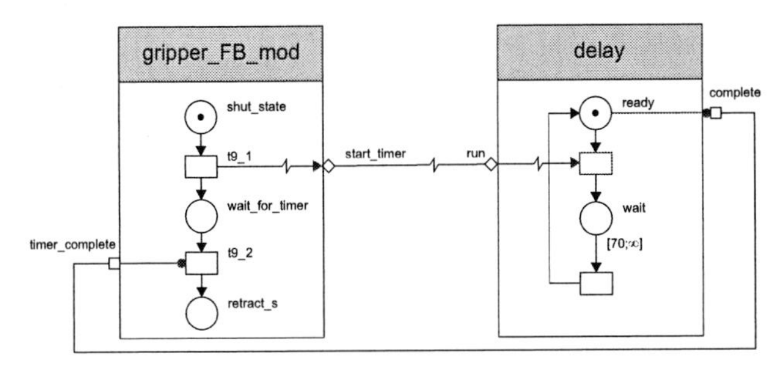

Figure 3.27.: Delay $_D$TNCEM.

to the values of the position sensors, either the left or the right branch is executed. To enter the left branch, the pallet has to be in handling position 1. The gripper is lowered (*extended_s*), the workpiece is gripped (*shut_state*), the gripper is raised (*retract_s*), and the conveyor is switched on (*conv_on_s*) again. The right branch is analog except that in handling position 2 the tin is disposed on the pallet. Finally, the sequence finishes with a token on the place *conveyor_off*. Each time a value changes, the *InStChanged* event is triggered to request the execution module for an update.
So far, the controller logic is just implemented as a sequence of states. This view corresponds to the SFC dialect of the IEC 61131, which is comprehensible considering the algorithm of the actual implementation shown in Figure C.1. However, the internal variables and timers of the controller have not been considered yet. Having a look at Figure 3.26 again, the transition *t9* fires without any further condition. In the real plant, this leads to a functional problem. If the controller closes the gripper and immediately retracts it, the tin will possibly not be gripped properly. For this, a delay time of 70ms is implemented in the control algorithm to pay attention to the plant dynamics. Considering this delay also is possible in the formal model. To do so, a *delay* function block is implemented according to the IEC 61131 standard. Figure 3.27 shows the modification of the gripper module. For this, transition *t9* is extended. First of all, *t9_1* triggers the firing of the event *start_timer*. Then, the token within the *delay* module moves from place *ready* to *wait*. After the step delay of 70, transition *t10* fires and the token is on place *ready* again. The condition *timer_complete* is finally fulfilled, and *t9_2* fires in the next step. The included step delay is without concrete dimension unit. So, it is not interpreted as a real-time interval. However, this interpretation is a meaningful extension of the modeling formalism, which is addressable in the verification tool. Admittedly, this goes beyond the scope of this thesis.
Figure 3.28 shows the implementation of a further delay timer that features an input condition. Again, the module will be started if the *run* event fires. Additionally, the condition *input* has to be fulfilled. After firing the transition *t1*, the token is on place

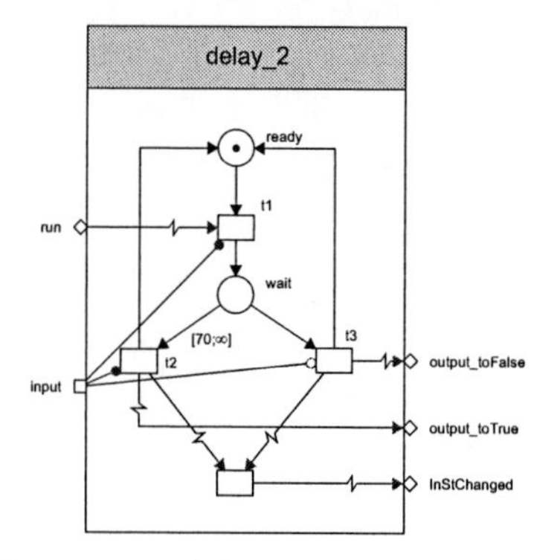

Figure 3.28.: A further delay $_D$TNCEM.

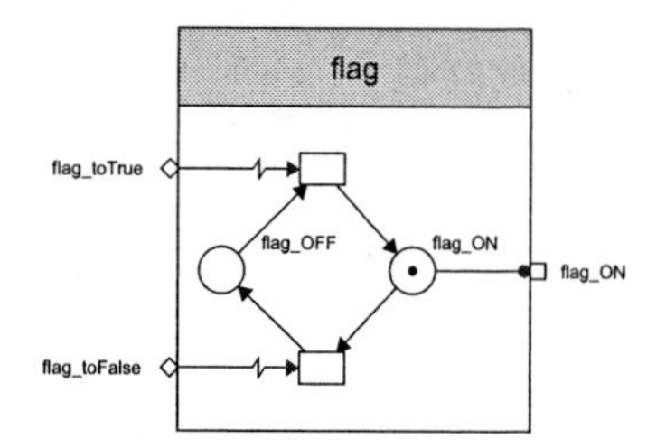

Figure 3.29.: Flag $_D$TNCEM.

wait for the step delay. Afterwards, the transition *t2* fires and the event *output_toTrue* is triggered. Finally, the token goes back to place *ready*. If the condition *input* gets false before, transition *t3* fires and the event *output_toFalse* will be triggered. An internal variable, the so-called *flag*, is modeled as shown in Figure 3.29. Its state transitions are triggered by incoming events and its state information is provided by means of a condition output. Of course, more complex functions can be implemented with $_D$TNCES as well. In [Ger11], the modeling of integer-valued data types as well as of Boolean and arithmetic functions is presented. In the scope of this work, however, the depicted functionality is sufficient.

In practice, the advantages of the IEC 61131 standard according to object-orientation and hierarchical implementation usually are not fully exhausted. It is not uncommon to write control software in a fairly large main program, which is hard to maintain. The structured modeling, which is proposed in this section, applies the possibilities given by the standard and encapsulates execution behavior as well as controller logic into functional units or modules, respectively. Nevertheless, such a detailed modeling is not mandatory since the whole controller could be packed into one single $_D$TNCES. However, it is the opinion of the author that control software engineering would benefit from the consequent application of standards and guidelines.

3.3.2. Model Generation

Manual controller modeling serves well for a plant that is designed from scratch to benefit from formal analysis as well as from an unambiguous behavior description of the target control software. Beyond this, the actual control code is better derived from the formal model. Nevertheless, this thesis is also motivated by the question how to support the migration of already existing controllers. Modernization is of rising importance in manufacturing industry because of expiring technical support for outdated controllers and raised demands according to computation power, flexibility, and adaptability. However, fragmentary documentation and poorly-commented source code not uncommonly lead to a complete reengineering of control software. Even though several functions might be portable, it is not assured that they behave as expected on the new hardware. Fortunately, the community early agreed on the IEC 61131-3 standard for the implementation of PLC software. This supports the automatic translation of IEC 61131-3-conform control code to formal $_D$TNCES and to further apply analysis. The workgroup, the author of this thesis belongs to, has started to deal with this topic in 1997 [HTLW97]. The authors present an approach to map function blocks to corresponding formal NCEM. The implementation is rather fixed and is able to handle only a subset of code structures. The ideas were taken up again in [GPH10] and developed further.

Formal controller models could be generated by mapping controller functions to $_D$TNCEM. However, the control code is based on a language. This implies that there would have to be one module for each possible and valid expression of this language. For example, a simple conjunctive combination of two or more variables needs one module per any possible cardinality of an input set. For this, mapping would need gigantic libraries of controller $_D$TNCEM. Nevertheless, especially this programming language is the key to an automatic translation. To do so, the algorithms have to be syntactically analyzed and compiled to $_D$TNCE controller modules. The actual implementation of a control code compiler is not performed in the scope of this thesis. For this, the reader is referred to [GPH10]. In the following, the principle is described. As a frame, the controller model in Figure 3.23 is applied again. The derived controller modules are embedded into the task module. Therefore, the control algorithm is divided into a control path, which is inserted to the *program* module, and into function modules, which are called by the *program* module. In the IEC 61131, different programming languages are defined. As it is specified in the standard, all programming languages can be converted to IL. Because of this, it is appropriate as a basis for the automatic code translation.

Algorithm 3.1 shows a cutout of the IL control code of the Gripper Station. It corresponds to the left branch of the controller module in Figure 3.26. The code is arranged in terms of so-called networks. Therefore, five function blocks are derived. The networks are executed sequentially and the resulting output values are stored to the process image. Due to the execution semantics, an output value can be overwritten by another function later in the network sequence. To be able to store internal or interim values, flags are included. Admittedly, this makes the control code hard to understand. Nevertheless, the application of IL as programming language is state-of-the-art. In the first network from line 1 to 6, four inputs are conjunctively combined. If the evaluation returns true, the output *extend_gripper* will be set to true and the output *retract_gripper* will be reset to false.

Algorithm 3.1: Gripper algorithm in IL.

```
1  LD      retracted                          //gripper retracted
2  ANDN    extended
3  AND     pos1_ON                            //pallet in handling position 1
4  ANDN    pos2_ON
5  ST      extend                             //extend gripper
6  R       retract
7
8  ANDN    retracted
9  AND     extended                           //gripper extended
10 AND     pos1_ON
11 ANDN    pos2_ON
12 ST      shut                               //shut gripper
13 ST      flag_1                             //set flag_1
14
15 delay(IN := flag_1, PT := T#70ms, Q => flag_2)
16                                            //wait for 70ms; afterwards, set flag_2
17 LD      flag_2                             //time elapsed
18 ANDN    retracted
19 AND     extended
20 AND     pos1_ON
21 ANDN    pos2_ON
22 R       extend
23 ST      retract                            //retract gripper
24 R       flag_1                             //reset flag_1 (flag_2 is reset meanwhile)
25 ST      flag_3                             //set flag_3
26
27 LD      flag_3
28 AND     retracted                          //gripper retracted
29 ANDN    extended
30 AND     pos1_ON
31 ANDN    pos2_ON
32 ST      conveyor_toON                      //start conveyor
```

That means the gripper is lowered. The delay function in line 15 will set *flag_2* to true if *flag_1* is true and the time of 70ms has elapsed. The value of *flag_2* remains true until *flag_1* is false again.

Figure 3.30 demonstrates the *task* module, which is derived from the IL code. The *program* module is displayed in detail. Again, it contains the event-driven execution sequence of the function blocks *FB_1* to *FB_5*. Underneath, the *FB_1* module is shown. If a *run* event is triggered, it will be evaluated whether the conditions *retracted* and *pos1_ON* are fulfilled, and whether *extended* and *pos2_ON* are not fulfilled. If so, transition *t1* will fire and will trigger the events *InStChanged*, *extend_toTrue*, and *retract_toFalse*. The function block modules *FB_2*, *FB_4*, and *FB_5* are constructed likewise. The internal nets of the *FB_3* and the *flag* modules are depicted in Figure 3.28 and 3.29, respectively. Finally, the *merge* module contains one transition in event mode $\boxed{\vee}$, which triggers the outgoing event each time an incoming event is fired.

The example depicts the idea of the automatic controller model generation. To do so, a set of rules has to be defined to parse the particular networks of the code and to derive the $_D$TNCEM. Thereby, the frame of the controller model is given as described in this section. In the author's workgroup, a set of rules has been developed. The actual implementation of a compiling tool is straightforward but still an open issue, which will have to be approached.

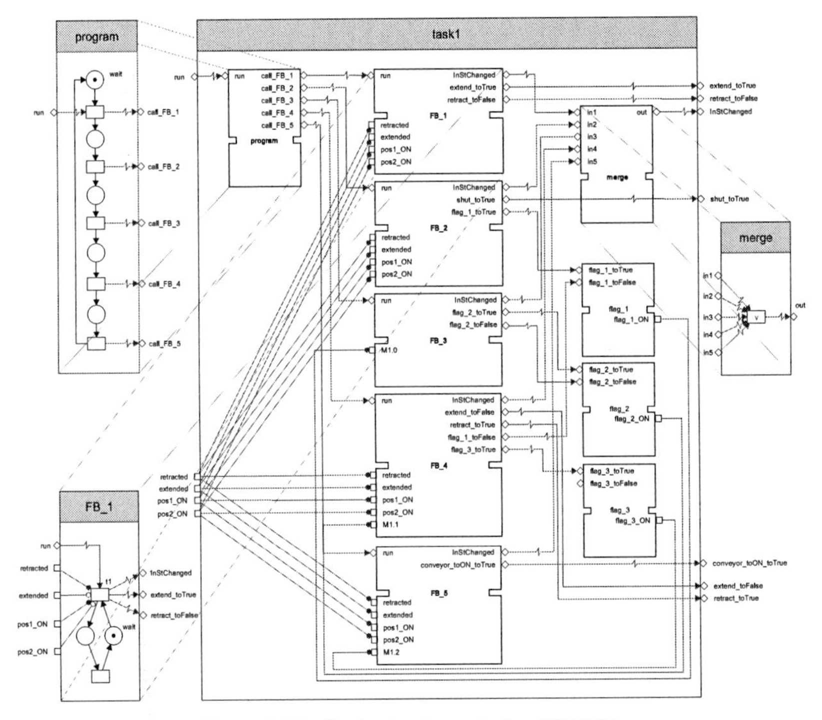

Figure 3.30.: Derived gripper task $_D$TNCEM.

3.4. Formal Closed-Loop Modeling

The previous sections present formal plant and controller modeling. Finally, both models are interconnected to establish the closed-loop system model in Figure 3.31. According to the signal concept of Section 2.6, condition edges are applied to link in- and outputs. In addition, one event connection is drawn to recognize state transitions of the sensors within the plant model. However, the author again would like to emphasize that this connection is not present in the real plant. It is just a means to handle the state space explosion, which would occur if the cyclic execution semantics of the controller was transferred one-to-one to the formal model. The interconnection of plant and controller model is done automatically by the verification tool, which is implemented in the scope of this thesis. For this, the condition outputs are linked to the condition inputs by means of their names. Accordingly, the event connection is

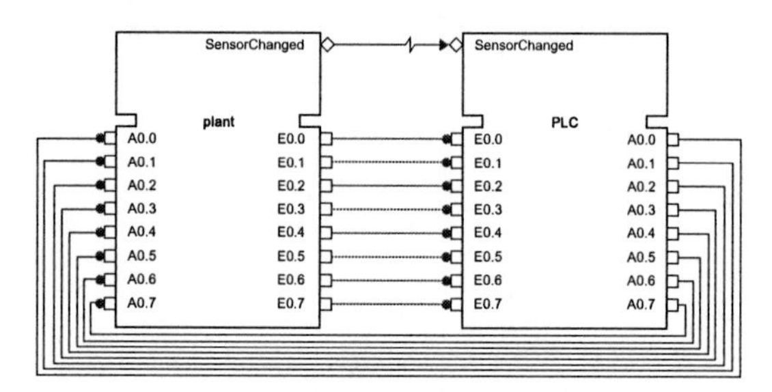

Figure 3.31.: Closed-loop ${}_D$TNCES.

drawn. The formal model of the closed loop is further applied to perform verification. To do so, a reachability analysis is performed and a formal specification of behavior is applied to it.

3.5. Summary

In this chapter, ${}_D$TNCES are applied to develop the formal models of the plant, the controller, and the closed-loop system. Regarding the plant, a concrete manufacturing system in lab-scale is considered. Special attention is paid to the component-based and hierarchical modeling since all modules are reusable in other plant configurations, too. ${}_D$TNCES provide a means to describe discrete time intervals. This potential is applied in a meaningful context. Beyond this, the capability of being integrated to the day-to-day plant engineering is in focus. As a first step, a methodology to derive a plant simulation from already existing design data is presented. Then, this simulation model is further used to generate a formal plant model. The tool, which handles this task, is developed in the scope of this thesis.
Concerning the controller model, special attention is turned to the modeling of the cyclic execution behavior. For this, the execution semantics and control algorithms as well as internal variables and timers are represented in the model. To support the application, the rules-based model generation out of control code is depicted.
Finally, the interconnection of plant and controller model to the closed-loop system model according to Section 2.6 is presented. Whereas the formal plant models serve well for documentation purposes because they are well defined and unambiguous, the formal controller models are further usable for control code generation. The main concern of this work, however, is the verification of the closed-loop system model. To do so, a formal specification of behavior is inevitable. This topic is in focus in the subsequent chapter.

4. Specification

A specification is the description of a product, a system or a service. Its aim is to define characteristics and attributes and to explain functionalities. A specification of a technical system shall be well defined, unambiguous, and comprehensible so that misunderstandings are avoided from beginning on. This is of crucial importance in daily engineering practice because the description of plant behavior is the very first step while designing a manufacturing system. For this, multiple interpretations of it must not be possible. Based on the description, the plant is constructed and the control software is implemented. Beyond this, it is essential for the exchange of ideas, the documentation, the subsequent modernization and migration, and last, not least for the analysis of plant and controller models. Especially the last point, namely the verification of models, is very important regarding safety because in contrast to simulation, verification considers every possible scenario and delivers a mathematical proof of correctness. However, these formal methods require expertise, which is not self-evident in daily practice.

The formal specification of plant behavior has become an important research topic as its mathematical basis allows analyzing the system with well-established methods and provides an unambiguous description of the system behavior. However, a specification is non-executable because it describes what the system does but not necessarily how. The automatic transformation of a system specification and a formal plant model to a controller implementation is called synthesis. For more information to this topic, the reader is referred to [Sch08, Mis12].

The application of intuitive but formal description techniques is motivated by the international standard IEC 61508 [IEC10]. The standard consists of seven parts and classifies safety-critical systems into four categories, whereas *Safety Integrity Level* (SIL) 1 is the lowest level and SIL 4 the highest one. The latest version was published in 2010. The IEC 61508 recommends the application of formal methods at least for the software part of a system. Due to reasons of traceability, it further demands an understandable explanation of formally-given requirements, which is essential for the documentation of the system. However, the explanation has to be unambiguous to exclude further sources for misunderstandings. Because of this, there is a need for techniques that support the description of formal requirements on the one hand, but that are comprehensible to engineers on the other hand.

These system requirements are divided into *functional* and *non-functional* ones. Functional requirements specify the plant behavior in terms of production requirements,

whereas non-functional requirements specify plant properties regarding for example safety, liveliness or the absence of deadlocks.
Plant behavior is specified with temporal logic formulas in this thesis. The formulas are very expressive and suitable to describe complex system behavior. Beyond this, they serve as an input language for the model checking tool. However, these nested expressions get harder to understand the more information is included. Because of this, errors may creep in during the development of the specification. Having in mind the adaptability of the proposed framework to daily practice, it will not be applied if it demands engineers to learn a completely new approach with an odd theory. So, there is a need for kind of front ends that enable engineers to create a specification without necessarily having to know all the details about the verification methods. For this, two domain-specific description technologies are proposed in this thesis. Both derive temporal logic formulas from more intuitive representations. Thereby, the application of a text-based approach as well as of a graphical one enables to cover the full spectrum of behavior requirements. The following sections introduce the two specification techniques of this thesis.

4.1. Safety-Oriented Technical Language

Text-based specification techniques are suitable especially to describe non-functional requirements. Usually, the constructs contain a small number of variables and specify desired and forbidden behavior. Using natural language is the most intuitive variant. Though, a translation to a non-ambiguous specification is not possible without restrictions. In contrast, a standardized language has a clear vocabulary and a formally-defined grammar. Usually, it is provided as a domain-specific language and applies phrases and terms, which the corresponding community is familiar with. Consequently, it has to be intuitive since the user must not be overcharged with theory. For this, the expressions have to be close to the language engineers use in practice. Going ahead, the formal specification is usable for the technical documentation as well because it describes the system behavior in an unambiguous way. However, the main advantage of a formal grammar - in the context of this work - is the ability to translate the expressions to temporal logic formulas, which are a means for the verification in terms of model checking.
The language, which provides these capabilities and which is developed in the scope of this thesis, is called Safety-Oriented Technical Language (SOTL). A first SOTL was introduced by a group of authors at TU Cottbus [HMD01] within the framework of the DFG project *Certifiable Logic Programmable Controllers* (CLPC)[1]. They specify requirements on control software of PLCs by using 18 sentences in the form of fixed

[1] *CLPC* project website. URL, http://www-dssz.informatik.tu-cottbus.de/DSSZ/Projects/CLPC, November 2013.

frames. These frames are completed with specific phrases, which correspond to variables and values. The language is focused on the specification of software behavior of IEC 61131-3-conform PLCs. For this, it considers the characteristic cyclic execution behavior. However, it is the plant behavior, which has to be specified for the application of this thesis' framework. The controller is just a means to operate the manufacturing system in an appropriate way. Consequently, it seems to be natural to describe and to check the desired plant behavior while analyzing the closed-loop system. For this reason, the SOTL is re-engineered in the scope of this work. The focus is now on the description of the desired plant behavior by considering the actual plant components. The language consists of 21 pattern phrases that can be nested arbitrarily. The formal grammar enables parsing and translating the expressions to CTL formulas. In the following, the syntax and semantics are depicted.

4.1.1. Syntax

The SOTL phrases contain certain keywords, which correspond to CTL expressions. A CTL formula is a combination of a path quantifier and a temporal operator. Accordingly, SOTL is constructed with modal and temporal terms. The modal terms are to be interpreted as stated below:

- A condition will hold necessarily if it is fulfilled on every path.
- A condition will hold possibly if it is fulfilled at least *on one path.*
- Boolean combinations are expressed by *and* (conjunction), *or* (disjunction), and *if ... then* (implication).

The temporal terms are to be interpreted as follows:

- If no temporal term is given, the variable values will be considered within one state.
- If a property holds *always*, it will be valid in every state.
- *Next* refers to the state following immediately.
- *Finally* is used if a requirement is fulfilled in a future state. Information about the concrete occurrence of the state is not supplied.
- A condition holds *until* another condition is fulfilled. As a demand, the second condition has to be true in a future state necessarily.
- A condition holds *before* another condition is fulfilled. As a demand, the first condition has to be true in at least one previous state necessarily.

4.1.2. Semantics

The context-free grammar of SOTL is depicted in Figure 4.1. It is given in *Extended Backus-Naur Form* (EBNF) according to the ISO/IEC 14977 [ISO96] standard. The corresponding syntax diagram is shown in Figure B.1.

<SOTL>	=	<Expression> .
<Expression>	=	[not] <Expression>
	|	it holds <Temporal> : <Expression>
	|	<Expression> <Conjunction> <Expression>
	|	<Expression> holds [on one path] , (until | before)
		<Expression> holds
	|	if <Expression> holds , then it holds
		<Temporal> : <Expression>
	|	<Variable>
<Conjunction>	=	and | or
<Temporal>	=	never | always [on one path]
	|	finally [on one path] | next [on one path]
<Variable>	=	{? ASCII character ? - _}
		(*every ASCII character except space*)

Figure 4.1.: SOTL grammar in EBNF.

Based on the grammar, the 21 basic SOTL constructs are derived as listed in the following. The terms φ and ψ represent expressions in the context of the grammar.

1. φ and ψ.
2. φ or ψ.
3. It holds always: φ.
4. It holds never: φ.
5. It holds always on one path: φ.
6. It holds next: φ.
7. It holds next on one path: φ.
8. It holds finally: φ.
9. It holds finally on one path: φ.
10. φ holds, until ψ holds.
11. φ holds on one path, until ψ holds.
12. φ holds, before ψ holds.
13. φ holds on one path, before ψ holds.
14. If φ holds, then it holds: ψ.
15. If φ holds, then it holds always: ψ.
16. If φ holds, then it holds never: ψ.

17. If φ holds, then it holds next: ψ.
18. If φ holds, then it holds finally: ψ.
19. If φ holds, then it holds always on one path: ψ.
20. If φ holds, then it holds next on one path: ψ.
21. If φ holds, then it holds finally on one path: ψ.

4.1.3. Compiler for SOTL

Since the language shall be applicable in practice, the *Compiler for SOTL* is implemented in the scope of this work. Figure 4.2 shows the graphical user interface (GUI). It supports the English version of SOTL, which is presented in this chapter. Furthermore, a German SOTL is implemented as well. The language is switchable through the menu option *language* of the compiler's GUI. The tool contains all grammar rules and automatically translates SOTL expressions to CTL formulas. For this, it performs a lexical analysis of the SOTL sentence. Then, it translates the derived structure to a CTL formula according to a set of replacement rules, which are given in Figure 4.1. CTL is chosen as a formal back-end to keep the language as simple as possible. The temporal logic offers enough expressive power to describe most non-functional requirements. However, an extension to derive eCTL and TCTL formulas is feasible, too. The software is written in C++ with the graphics package wxWidgets[2]. The included compiler is implemented in a C environment with the help of the tools Flex[3] and Bison[4].

The software features four different tabs. *Custom expression* includes the 21 pattern phrases of SOTL. The other tabs contain specific phrases for *safety expressions*, *liveliness expressions* and *deadlock-free expressions*. The patterns are chosen via drop-down

[2]wxWidgets. URL, http://www.wxwidgets.org, November 2013.
[3]Fast Lexical Analyzer (Flex). URL, http://flex.sourceforge.net, November 2013.
[4]Bison. URL, http://www.gnu.org/software/bison, November 2013.

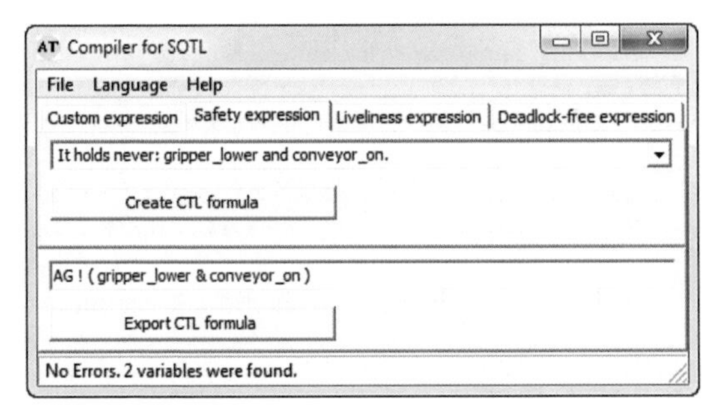

Figure 4.2.: Compiler for SOTL.

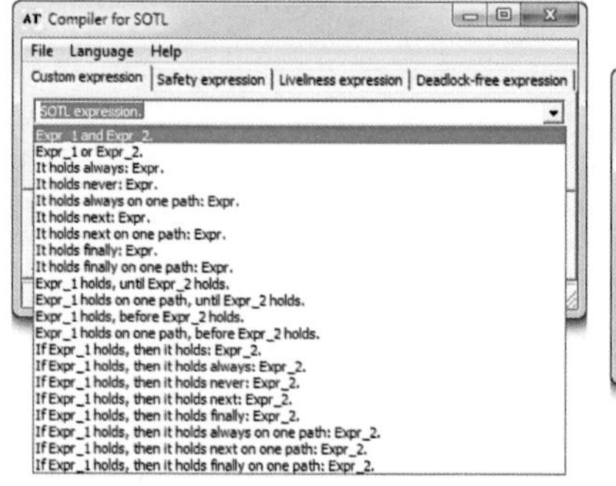

Figure 4.3.: Custom patters.

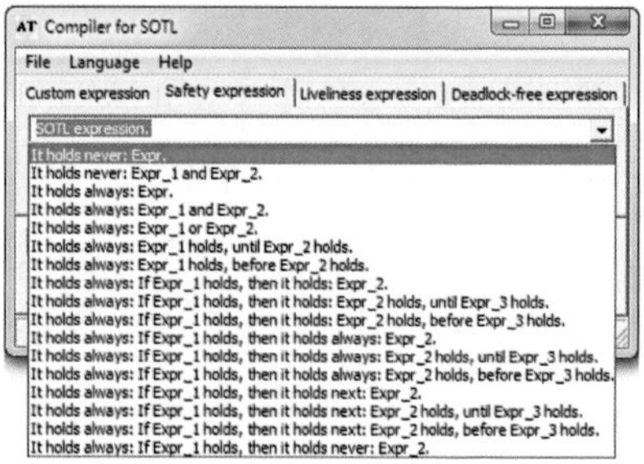

Figure 4.4.: Safety patterns.

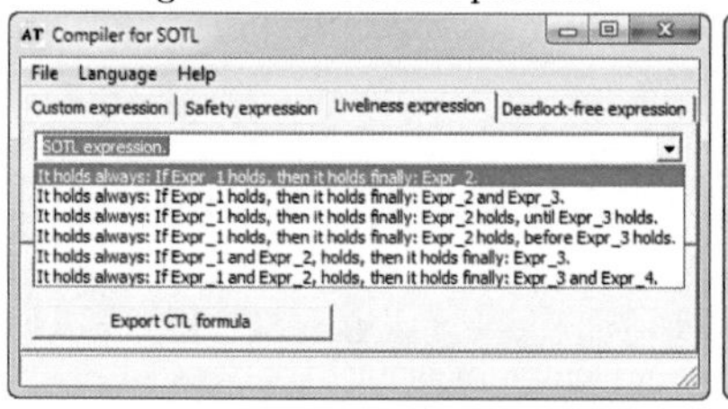

Figure 4.5.: Liveliness patterns.

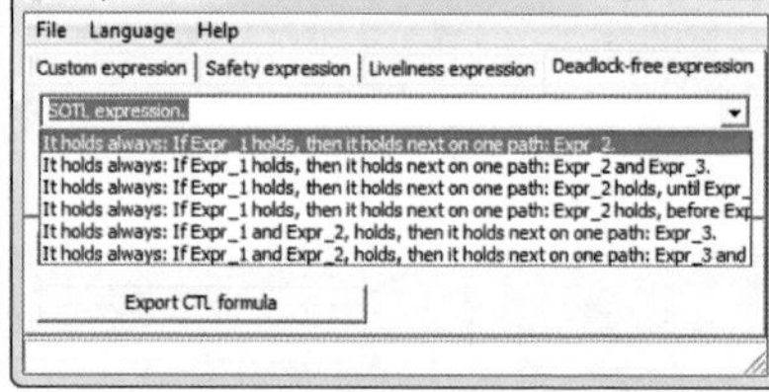

Figure 4.6.: Deadlock-free patterns.

menus, which are shown in Figure 4.3 to 4.6, and serve as assistance while developing the specification. However, the user is free to enter any custom expression. The compiler checks whether the input is valid according to the grammar rules. It derives and displays the CTL expression by clicking on *Create CTL formula.* In case of error detection, the software puts out a message at which position within the SOTL expression the error can be found. The derived formulas are exported to the model checking tool, which is developed within the framework of this thesis. Since they are provided in plain text, they are further usable for other applications, too.

To give some specification examples, the following non-functional requirements are considered with respect to the Gripper Station in Figure 3.3. As a safety property, it must not occur that the gripper of the demonstrator and the conveyor are both moving in the same state. This is expressed by Formula 4.1. Liveliness requirements specify desired behavior as shown in Formula 4.2. If the conveyor starts moving, the pallet will finally reach position 1 or position 2. The absence of deadlocks is checked through verifying that a possible action can be executed in at least one successor state. For example, if the pallet reaches position 2, the gripper will be lifted down, which is expressed by Formula 4.3.

The examples show that SOTL phrases are understandable in a technical specification since they are close to the technical language engineers apply. In addition, the formal grammar enables deriving CTL formulas.

SOTL: It holds never: gripper_lower and conveyor_on.
CTL: AG ! (gripper_lower & conveyor_on) (4.1)

SOTL: It holds always: If conveyor_on holds, then it holds finally: conveyor_pos1 or conveyor_pos2.
CTL: AG (conveyor_on $\rightarrow$ AF (conveyor_pos1 $\vee$ conveyor_pos2)) (4.2)

SOTL: It holds always: If conveyor_pos2 holds, then it holds next on one path: gripper_lower.
CTL: AG (conveyor_pos2 $\rightarrow$ EX $gripper\_lower$) (4.3)

The application of text-based specification technologies is limited by the amount of data, which shall be displayed. This arises while considering a whole production sequence with many variables and states. A lot of text would be necessary for the description. Because of this, graphical specification techniques are applied in addition. The next section introduces such a methodology.

4.2. Symbolic Timing Diagrams

Graphical specification techniques are applied if a lot of information shall be displayed in a well-arranged way. A possibility to describe complex program sequences and functional requirements, respectively, is supplied by the STDs. They are basically used in hardware development to describe the temporal behavior of reactive systems [SJW98]. Because of their expressiveness and due to their structure, STDs are used in this thesis for specifying plant behavior in an intuitive way. Referring to the diagram formalisms, which are presented in Section 2.5.4, some modifications and simplifications are conducted by the author so that STDs are applicable for the specification of production sequences. These modifications are discussed later in this section.
As shown in Figure 4.7, an STD consists of one or more waveforms, each one describing a discrete state sequence of a variable. State transitions can occur synchronously, which is shown by a solid blue vertical connecting line. An STD depicts the time lapse in horizontal direction by means of discrete time steps. The state transitions of two waveforms can be interrelated with a constraint that specifies the interval after which the target transition fires at the earliest and at the latest after the source transition has fired. A constraint is depicted as a solid green directed arrow with closed interval

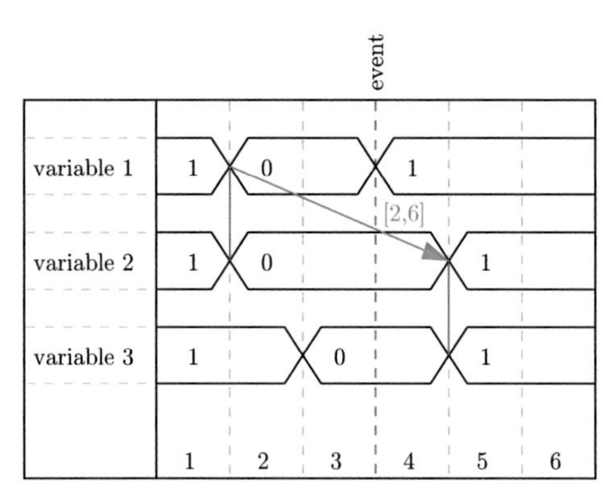

Figure 4.7.: Example for a Symbolic Timing Diagram.

information in squared brackets. State transitions can be specified by event information. Events are depicted with dashed red vertical lines and hold for all connected waveform transitions. According to the information within a diagram, temporal logic formulas are derived. This procedure is shown later on in this section. In the following, the modifications of STDs are discussed.

4.2.1. Modifications

The formalism of STD is comprehensive and facilitates many degrees of freedom to specify system's behavior. However, the claim of this thesis is to provide intuitive description methods, and because of this, the user shall not be overcharged with specification techniques that relocate the complex theory of temporal logics to the front ends of the corresponding software tools. For this, the applied graphical description forms should be as simple as possible to remain understandable for plant engineers. This is achieved by limiting the expressive capabilities.

To describe production requirements as a sequence of states, the waveforms are developed in discrete steps, each one specifying a certain plant state. Synchronous state transitions of two waveforms (displayed by a solid blue vertical line) qualitatively express that state z_{i+1} follows state z_i ($i \in \mathbb{N}_0$) without making an assertion about the time lapse. The interpretation corresponds to the asymmetric simultaneous qualitative constraints in Table 2.1. To quantitatively specify an interval between two state transitions, interval constraints can be added to the diagram (displayed by a solid green directed arrow). In contrast to the different interval annotations of constraints in Table 2.1, only asymmetricly-combined quantitative constraints are applied in this thesis. For this, the interval $[n, m]$ is valid for $n, m \in \mathbb{N}_0$ and $n \leq m$.

The state sequences are not divided into mandatory and possible ones. For this, a diagram will hold only if the complete specified sequence occurs in the dynamic graph of the system. Furthermore, it is not distinguished whether a diagram holds initially or in a future state.
The modifications ease the automatic translation of an STD to a temporal logic formula, which is addressed next.

4.2.2. Translation

The translation of timing diagrams to temporal logic formulas is still a matter of research for different workgroups. On the one hand, timing diagrams offer a means to specify sequences in a descriptive way, and on the other hand, temporal logic formulas are a practicable input format for model checking tools. In this thesis, CTL, eCTL, and TCTL are applied. For this, the translation of the STDs in Figure 4.8 to 4.10 to these temporal logics is described in the following.
To derive a CTL formula, the state information of every variable or waveform, respectively, is extracted for each discrete diagram state. Having a look at Figure 4.8, the values of all three variables are true in the first state. Accordingly, all variables are true in the first specified state of the CTL Formula 4.4. A diagram holds in at least one post-state, which is expressed by the term EF. The partial CTL formulas of the subsequent diagram states are conjunctively linked. Thereby, each following diagram state is added in connection with the term EF. That means the diagram state z_{i+k} has to be true in one post-state on at least one trajectory through the closed-loop system's state space starting in state z_i $(i, k \in \mathbb{N}_0, k > 0)$. Doing so, the following structure is built for the five diagram states of the STD in Figure 4.8:

$$\text{EF (state 1 \& EF (state 2 \& EF (state 3 \& EF (state 4 \& EF (state 5))))).}$$

Since the STD does not contain all system variables necessarily, there could be states and state transitions in the controlled plant model that are not captured in the diagram. Therefore, the diagram states represent a subset of the closed-loop system states, and because of this, the operator EF and not EX is applied. Consequently, the derived formula is weak and not sufficient to specify non-functional requirements such as safety. Even though the derived Formula 4.4 looks expanded, it is essential to specify the value of every variable in each state. This is because an STD shows the pre- and the post-states of one particular state. In contrast, a CTL formula only expresses the properties of the post-states. Nevertheless, the conjunctive combination of all variable values in one state affects the model checking runtime only linearly since no additional temporal correlations have to be checked. Thus, the resulting CTL structure in fact is large but simple. In contrast, specifying each waveform with one partial formula and then interconnecting these formulas according to their temporal relations as proposed in [CF05], would relocate the complexity to the specification of

the state transitions. This would lead to a CTL formula, which is more complicated to analyze for the model checking tool.
CTL is suitable to describe the states of a system, but however, it is not capable to specify state transitions explicitly. To extend CTL in that way that state transition information can be specified, eCTL has been proposed. Formula 4.5 shows the eCTL formula, which corresponds to Figure 4.9. As CTL is a subset of eCTL, the expression is analog to Formula 4.4. It is extended by the state transition information from diagram state 1 to 2. That means that the *event* has to occur to enable the transition from diagram state 1 to 2. For this, the formula has the following structure:

EF (state 1 & E event F (state 2 & EF (state 3 & EF (state 4 & EF (state 5))))).

With CTL, qualitative temporal relations are specified. However, no assertions regarding time intervals can be expressed. For this, TCTL has been proposed to define the time interval between two state transitions explicitly. Thereby, the concrete time unit is not given. Formula 4.6 shows the TCTL formula, which corresponds to Figure 4.10. Again, CTL is a subset of TCTL, and because of this, the expression is analog to Formula 4.4. The time interval is appended to the CTL formula by the operator EF. The source state and the target state are added and combined as follows:

EF (state 1 & EF (state 2 & EF (state 3 & EF (state 4 & EF (state 5))))) &
EF (state 1 & EF [6,10] (state 5)).

A further interval would be appended in the same way by adding the operator EF and then specifying the interval constraints. To support the specification with STD and to generate the corresponding formulas automatically, an editor is implemented in the scope of this work and presented subsequently.

4.2.3. STD Editor

For the application in practice, the *STD Editor* in Figure 4.11 is implemented. The tool is written in C++ applying the graphics package wxWidgets. A diagram is developed in discrete steps. For this, the two buttons *Step +* and *Step -* are featured to append new states and to delete existing ones. Thereby, the variable values are switched by radio buttons for true and false. The editor handles an arbitrary quantity of variables and states. Besides the standard functions to load and save, a diagram is exported to pictures in *.png or *.eps file format. For example, the diagrams in Figure 4.8 to 4.10 are designed and exported with the STD Editor. An existing diagram is edited by right-clicking on the corresponding region. As visible in the diagram cutout in Figure 4.12, the variable values, the event information, and the temporal intervals are changeable through a drop down menu. Clicking on *Change variable value* generates the resulting diagram in Figure 4.13. An event is similarly

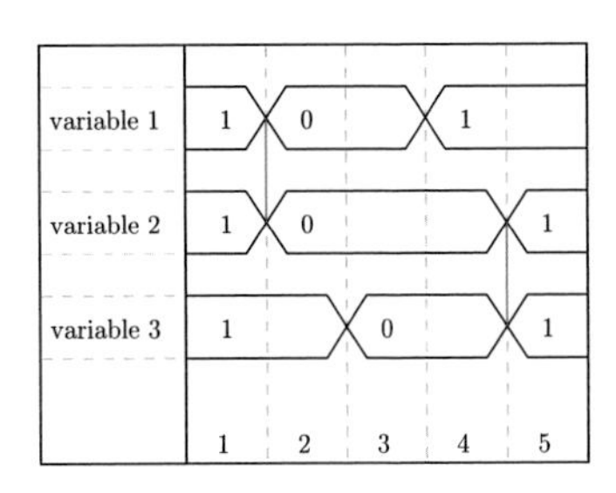

Figure 4.8.: STD to derive a CTL formula.

EF (variable 1 & variable 2 & variable 3 & *EF* (! variable 1 & ! variable 2 & variable 3 & *EF* (! variable 1 & ! variable 2 & ! variable 3 & *EF* (variable 1 & ! variable 2 & ! variable 3 & *EF* (variable 1 & variable 2 & variable 3))))) (4.4)

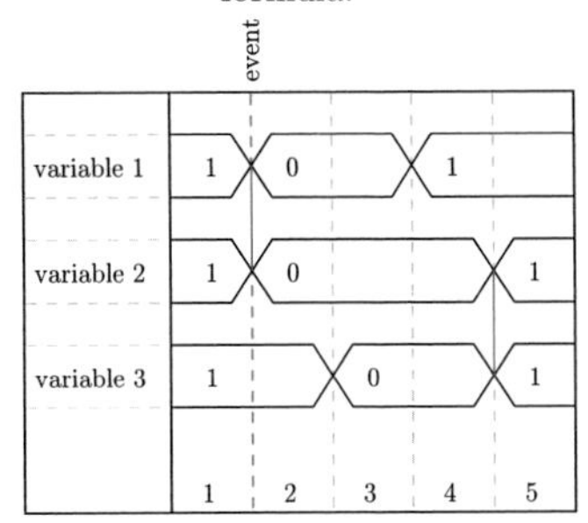

Figure 4.9.: STD to derive an eCTL formula.

EF (variable 1 & variable 2 & variable 3 & *E event F* & (! variable 1 ! variable 2 & variable 3 & *EF*(! variable 1 & ! variable 2 & ! variable 3 & *EF* (variable 1 &! variable 2 & ! variable 3 & *EF* (variable 1 & variable 2 & variable 3))))) (4.5)

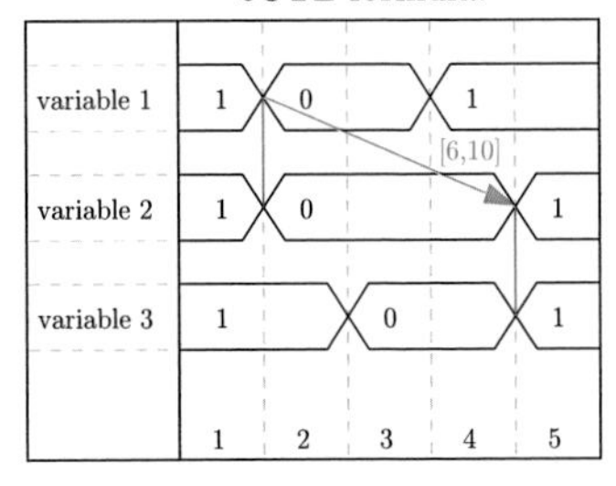

Figure 4.10.: STD to derive a TCTL formula.

EF (variable 1 & variable 2 & variable 3 & *EF* (! variable 1 & ! variable 2 & variable 3 & *EF* (! variable 1 & ! variable 2 & ! variable 3 & *EF* (variable 1 & ! variable 2 & ! variable 3 & *EF* (variable 1 & variable 2 & variable 3))))) & *EF* (! variable 1 & ! variable 2 & variable 3 & *EF* $[6, 10]$ (variable 1 & variable 2 & variable 3)) (4.6)

added as shown in Figure 4.14. To include an interval, an arrow is drawn from one waveform transition to another. The numbers for the lower and upper interval bounds

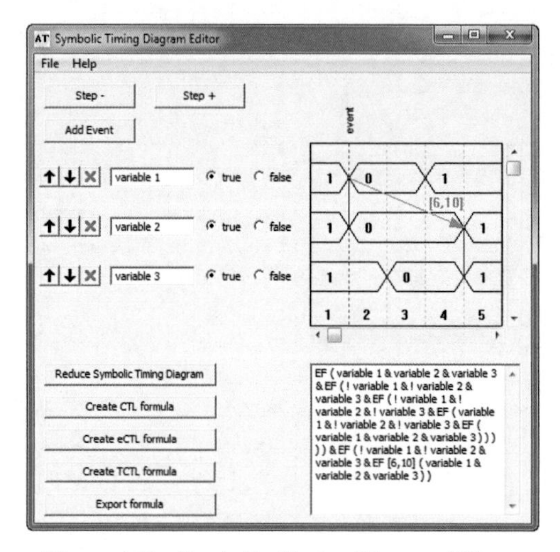

Figure 4.11.: Symbolic Timing Diagram Editor.

are defined in a dialog as depicted in Figure 4.15. Thereby, -2 is interpreted as ∞ to mark the upper bound as infinite. Redundant diagram information is automatically reduced by combining identical successive states. Of course, events and intervals are considered so that no information is deleted by mistake by clicking on the button *Reduce Symbolic Timing Diagram.* Doing so, the derived temporal logic formulas are minimal according to the translation procedure, which is described in the previous subsection. A derived CTL, eTCL, or TCTL formula is displayed in a text box and exported to the model checking tool.
To show the application, again the Gripper Station, which is shown in Figure 3.3, is considered. The STD in Figure 4.16 specifies the gripper sequence. The corresponding variables are the actuators *gripper_lower* and *conveyor_on* as well as the sensors *conveyor_pos1* and *conveyor_pos2*. The sequence starts with all variable values on false. After the event *ev_start* fires, the conveyor starts moving. If the pallet on it reaches the first position, the conveyor will be stopped. After the event *ev_lower* has fired, the gripper goes down and up again. Then, the pallet moves to the second position and the gripper is lowered and raised again. Finally, the conveyor transports the pallet to the next procession station and both sensors are deactivated. In addition, it is specified that the conveyor stops within the discrete interval $[0, 5]$ after sensor *conveyor_pos1* or *conveyor_pos2* has been activated. The following formulas display cutouts of the CTL (Formula 4.7), eCTL (Formula 4.8) and TCTL (Formula 4.9) formulas, which are derived from the STD given in Figure 4.16.

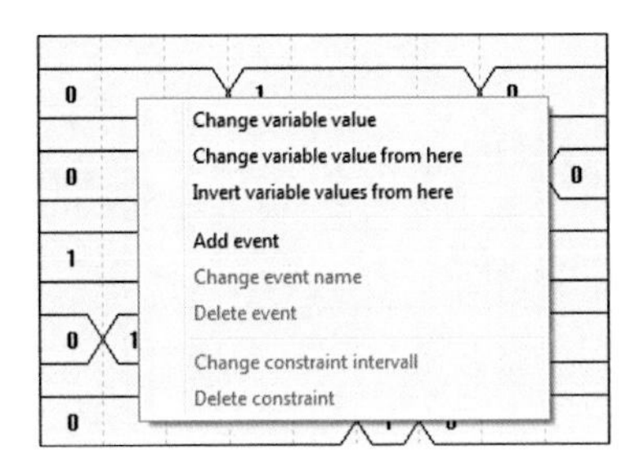

Figure 4.12.: Drop-down menu.

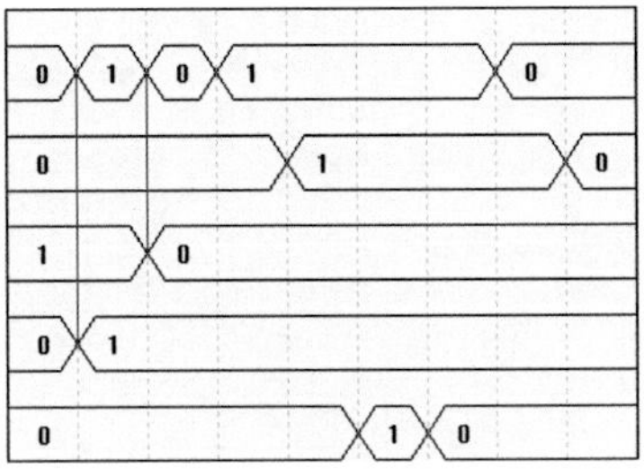

Figure 4.13.: Changed variable value.

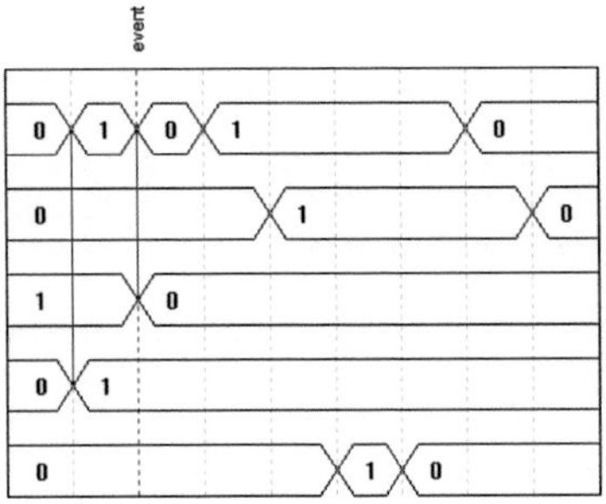

Figure 4.14.: Added event.

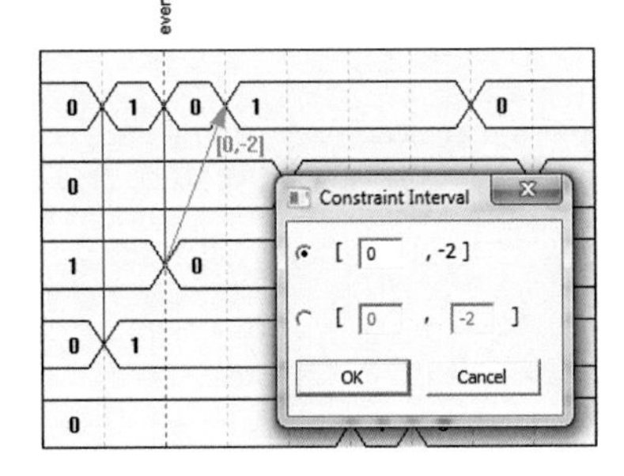

Figure 4.15.: Added temporal interval.

EF (! gripper_lower & ! conveyor_on & ! conveyor_pos1 & ! conveyor_pos2
&EF (! gripper_lower & conveyor_on &! conveyor_pos1 & ! conveyor_pos2
& EF (...))) (4.7)

EF (! gripper_lower & ! conveyor_on & ! conveyor_pos1 & ! conveyor_pos2
& E *ev_start* F (! gripper_lower & conveyor_on & ! conveyor_pos1
& ! conveyor_pos2 & EF (...))) (4.8)

EF (! gripper_lower & ! conveyor_on & ! conveyor_pos1 & ! conveyor_pos2
& EF (! gripper_lower & conveyor_on & ! conveyor_pos1 & ! conveyor_pos2
& EF (...))) & EF (! gripper_lower &conveyor_on & conveyor_pos1
& ! conveyor_pos2 & EF $[0, 5]$ (! gripper_lower & ! conveyor_on
& conveyor_pos1 & ! conveyor_pos2)) & EF (...) (4.9)

An STD serves well for the specification of a production sequence. The export functionality of the presented tool allows including the diagram as well as the formula to a

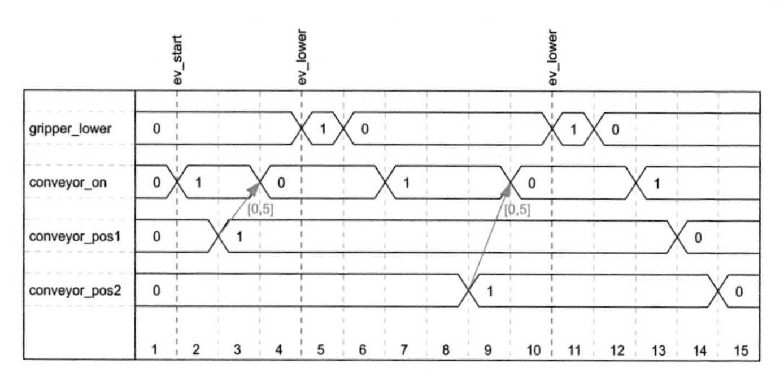

Figure 4.16.: Gripper sequence specified with an STD.

technical specification. The tool itself is extendable to include further information as well. For this, it contributes to the well-defined but also domain-specific description of plant behavior.

4.3. Summary

In this chapter, the formal specification of plant behavior is addressed. For this, temporal logics are applied to get a well-defined and unambiguous description of properties. However, these formulas are hard to synthesize. Beyond this, their complex theory complicates the application in daily plant engineering. Nevertheless, they serve as an input language for the model checking procedure. Consequently, their utilization is inevitable in the framework of this thesis.
To tackle the challenges, two meta-description technologies are proposed in this chapter. First of all, the SOTL provides a specification language based on a formal grammar. As supposed by its name, the constructs are applied especially to describe safety properties. On the other hand, STDs are chosen as a graphical approach because they serve well to display state sequences. Beyond this, engineers are familiar with timing diagrams as a domain-specific description tool. Because of the formal basis of the two approaches, translation rules to temporal logic formulas are defined. Beyond this, the methodologies are implemented in software tools, namely the Compiler for SOTL and the STD Editor. Both feature intuitive front ends and hide the theory in background. Furthermore, they automatically derive the formulas, which are used as an input for the model checking tool. The subsequent chapter applies all of this thesis' approaches and discusses how to study the closed-loop behavior.

5. Analysis of the Closed-Loop Behavior

In this chapter, the approaches of the previous ones are applied to study the behavior of the closed-loop system. Thereby, the focus is on simulation on the one hand, and on verification on the other hand. Although both are related, they differ in the applied technologies. For this, simulation is more or less an informal testing methodology, whereas verification delivers a formal proof of correctness. Subsequently, the simulation framework of this thesis is considered in detail. Afterwards, verification is taken into account.

5.1. Simulation in Closed Loop

Simulation is performed to study the behavior of a system. For this, a model of the real system is abstracted to run test cases on it. Thereby, the particular real system is usually not affected. There are several motivations to apply simulation techniques because different (scientific) domains have different views on it. Of course, simulation also is a well-established tool in many engineering areas, like for example material science or chemical engineering. However, it is a kind of novel approach in the plant engineering field because of its fairly conservative structures. Despite the numerous advantages, simulation also has some critical points to be mentioned. First of all, the model generation is connected to additional work expenses. A more crucial point is that the quality of a simulation is only as good as the quality of the model. That means that a simulation will be useless if the underlying model is too abstract to represent the real system or if it is even incorrect. Because of this, the correct model generation is vitally important.
As this thesis is especially focused on manufacturing systems control, it approaches the challenges of plant simulation. There are numerous (commercial) software tools that enable the simulation of continuous (for example The MathWorks®[1] Simulink, Chemstations^{TM}[2] CHEMCAD) and discrete processes (for example Incontrol[3] Enter-

[1]The MathWorks website. URL, http://www.mathworks.com/, November 2013.
[2]Chemstations. URL, http://www.chemstations.com/, November 2013.
[3]Incontrol website. URL, http://www.incontrolsim.com, November 2013.

prise Dynamics®, DUALIS®[4] SPEEDSIM, ITI[5] SimulationX®). Therefore, this work does not present a further simulation software implementation. It rather approaches the question whether the quality of simulation is sufficient to make a dependable assertion according to the correct implementation of control software. For this, existing approaches have to be integrated to the engineering workflow and extended with formal analysis as further described in this chapter.
In the following sections, HiL and SiL Simulation are discussed. As the name implies, the closed-loop of controller and plant is considered. Thereby, both technologies require a plant model. The main difference is that the control software runs on the target controller for HiL Simulation. In contrast, it runs in a simulated environment on a PC performing SiL Simulation. The specific advantages are depicted in the following.

5.1.1. Hardware-In-The-Loop Simulation

Usually, simulation is performed by applying test cases in terms of input sequences to the corresponding controller. The control device processes this input data according to the control algorithms and provides output information, which is evaluated regarding consistency. Doing this in an open loop is no real solution since the whole complexity of a plant, especially of tasks running concurrently, is not manually manageable. Because of this, the closed-loop simulation is a more powerful approach to test the correctness of software. To do so, test cases are provided by means of a simulated plant model, which contains a comprehensive model of plant behavior that is semi-automatically generated (see Section 3.2.2). For this, synthetic test cases are not applied to the controller, but the plant model receives controller output and provides controller input information as in a real system. Doing so, the correct software execution is directly traceable regarding the plant simulation.
For the HiL Simulation, there are no limitations according to the applied controller hardware as long as it features an interface to a PC. Because of this, it is vendor-independent and suitable for almost any kind of controller. The other way round, it makes no difference for the controller whether it controls the real plant or its simulation model. For this, HiL Simulation is very suitable to test the software under conditions, which are very close to reality, before starting up the real plant. To do so, the controller is connected through its inputs and outputs to the simulation model. In fact, this is a challenging task because the simulated signals of plant's sensors and actuators have to be transmitted to and from the control device. For this, additional hardware is necessary to interconnect the simulation and the controller periphery.
The experimental setup for the HiL Simulation, which is performed in this thesis, is shown in Figure 5.1. The hardware is composed of the components that are listed in Table 5.1. The controller is a Siemens S7-300 CPU ①, which belongs to one of

[4]DUALIS website. URL, http://www.dualis-it.de/, November 2013.
[5]ITI website. URL, http://www.iti.de/, November 2013.

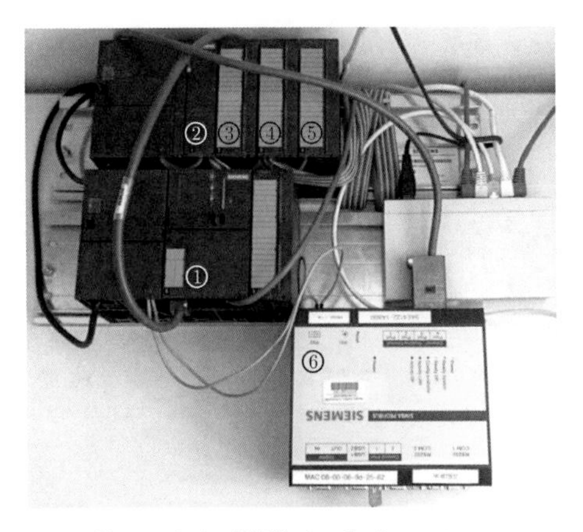

Figure 5.1.: PLC simulation setup.

#	device	catalog number	comment
①	SIMATIC S7-300 CPU315-2 PN/DP	315-2EH13-0AB0	PLC
②	SIMATIC ET200M	153-2BA02-0XB0	I/O station
③	SM321 DI 16 x DC 24V	321-1BH01-0AA0	digital input unit
④	SM322 DO 16 x DC 24V	322-1BH01-0AA0	digital output unit
⑤	SM334 AI 4 / AO 2 x 8 BIT	334-0CE01-0AA0	analog I/O unit
⑥	SIMBA Profibus PNIO	9AE4122-1AB00	simulation unit

Table 5.1.: Controller hardware system for benchmarks.

the mostly-applied control devices in industry. In practice, it is connected through Input/Output (I/O) units ③, ④, ⑤ to the sensors and actuators of the actual physical process. One possible configuration is to directly attach the units to the PLC. However, the cabling effort is fairly high considering huge plants because each wired field device needs a physical connection to the I/O unit. Consequently, another implementation is to combine all field signals, which belong to a specific plant part, and bring them together in one modular I/O station, namely an ET200M ②. Several of these ET200M devices are connected through a communication bus to the PLC, and the cabling effort is reduced to one physical bus wire.
The Siemens field bus simulation system applies the second approach. For this, the I/O station ② is simulated by the Siemens SIMBA PNIO unit ⑥ to be able to affect sensor and to observe actuator signals directly. This enables to connect the controller to a plant simulation. To do so, the purple bus wire is just unplugged from the I/O station and attached to the SIMBA device. The procedure is transparent for the con-

troller, which means the controller does not "realize" any difference between running the real plant and running its simulation.
The SIMBA unit features four field bus channels, each one capable to simulate 125 slaves, and offers Profibus and Ethernet communication links. Controller and SIMBA box are interconnected via Profibus, whereas simulation PC and box are connected via Ethernet. The exchange of data is performed through communication functions, which are provided by the SIMBA simulation system. Hence, the simulated plant receives actuator signals from and sends sensor signals to the controller. The communication functions of the SIMBA device are embeddable into simulation as well as into verification environments. Within the scope of the OMSIS project, this integration has been implemented for the simulation software tools Enterprise Dynamics®, DUALIS® SPEEDSIM and ITI SimulationX®. The actual involvement is less a challenge but more a routine piece of work for a software engineer. For this, it is especially the turn of soft- and hardware vendors to include the corresponding capabilities so that HiL Simulation can be performed.
The software engineering environment for Siemens PLCs, namely the *Simatic Step 7*, exports the PLC hardware configuration to the *SIMBApro* configuration tool displayed in Figure 5.2. Thus, the PLC inputs can be forced and the PLC outputs can be observed. Instead of performing such an open-loop test, the communication functions are embedded to the simulation environment IED. Doing so, the closed-loop HiL Simulation is established. Figure 5.3 displays the simulation of the demonstration plant example in IED. The whole production process is controllable, and simulation runs are executed without affecting the real plant. Thereby, dynamic aspects are considered in particular. However, there is a certain weak point. To identify it, the HiL Simulation scheme of this thesis is illustrated in Figure 5.4. The plant simulation model is interconnected with the controller hardware through the SIMBA device ①. This configuration is sufficient to test whether the specified production sequences are executed correctly. Nevertheless, each simulation run just regards one possible trajectory within the reachable state space. For this, several runs are necessary to capture all possibilities. Consequently, the elementary question arises: Is there an indicator for the completeness of the state space?
Of course, this question is a theoretical one because the indicator is the regarded state space itself. A possible way to approach this problem is to include formal analytic methods to the simulation process. For this, a $_D$TNCES plant model is derived from the simulation model as described in Section 3.2.2. This model is imported to the TMoC and visualized as shown in Figure 5.5. The right window part displays the states of the plant components. Green means that the corresponding place is labeled, whereas red means that it is unlabeled. The left window part shows the plant in- and output states. The interconnection of the controller outputs to the formal model inputs is done automatically. For this, the configuration is imported from the SIMBApro tool. The TMoC polls the controller outputs. Each time a change is recognized, the information is transferred to the plant model ②. Doing so, the places

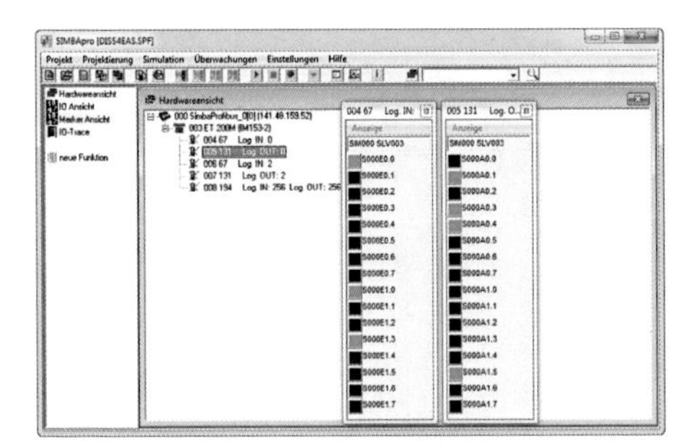

Figure 5.2.: SIMBApro configuration software.

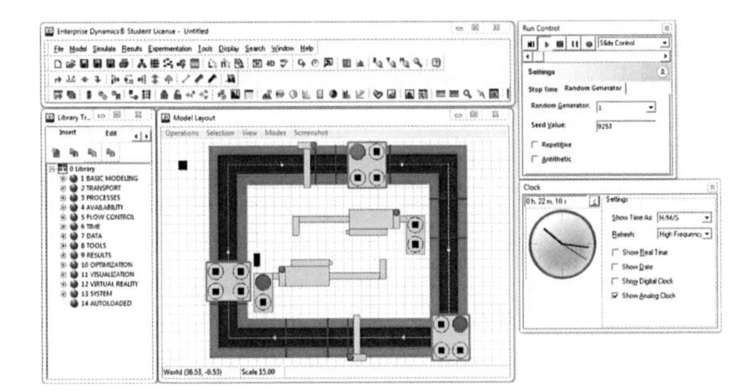

Figure 5.3.: IED simulation environment.

within the plant input modules are labeled accordingly. In combination with the state of the formal plant model, an initial marking is given. The TMoC calculates all successor states ③ based on this initial marking. Doing so, a sub-graph is computed for any new controller output information. The final states of one sub-graph are the *branching states*. Analytically, these states are dead because they have no post-state. For this, they are interpreted as states, in which the formal plant model expects new controller information to proceed. The dynamic graph is composed by interrelating one branching state and the initial state of the next sub-graph. For this purpose, the branching state is chosen by comparing the plant simulation model outputs, which are the sensors, to these ones of the formal plant model ④. The analysis continues only if the data is consistent in one state at least. Otherwise, a modeling error is revealed. Doing so, a closed trajectory from the very first state to the final plant state is derived. This procedure is performed concurrently to the plant simulation in IED. As an example, Figure 5.6 shows the state space of the first complete HiL Simulation run of the Gripper Station in Figure 3.3. This graph is produced by the TMoC. The

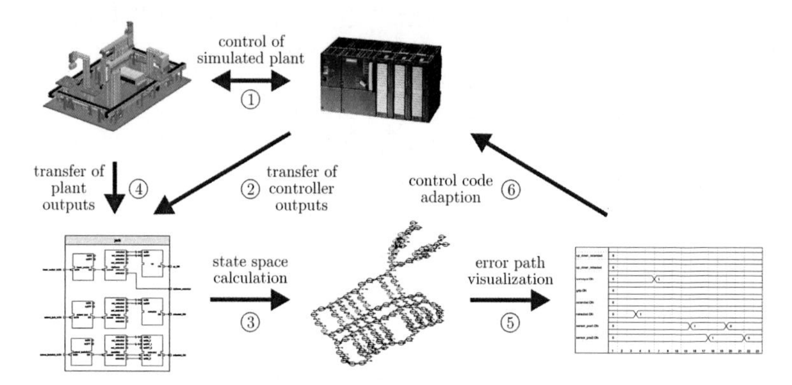

Figure 5.4.: HiL Simulation scheme.

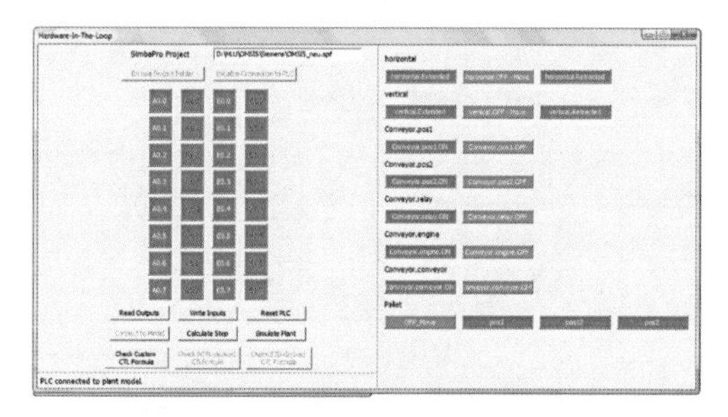

Figure 5.5.: TNCES Model Checker.

initial state is the furthest left, whereas the final state is the furthest right. Based on the controller output information, the TMoC computes the 60 states colored in black and green. These 60 states mark the closed trajectory from the first state to the last one, which means that the tin on the pallet has been processed correctly. Thereby, the green states mark the 8 state transitions from a branching state to an initial state of an adjacent partial graph. For this, these transitions are triggered by a change of the controller outputs. Since the process is finished in a valid final state, this individual simulation run is performed successfully. However, the TMoC has calculated 31 additional branching states, which are colored in red. To have a comprehensive simulation, these branching states also have to be considered by performing more simulation runs. Thereby, the TMoC finds existing states and combines their pre- and post-relations so that already-considered control branches are captured and open trajectories are closed. Accordingly, the simulation will be complete if there are no

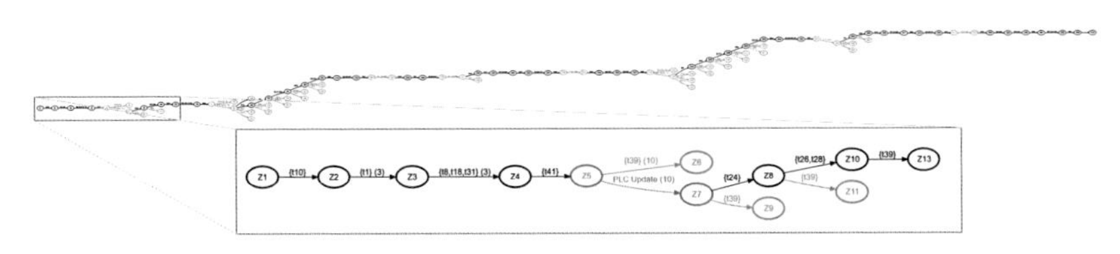

Figure 5.6.: State space of one simulation run.

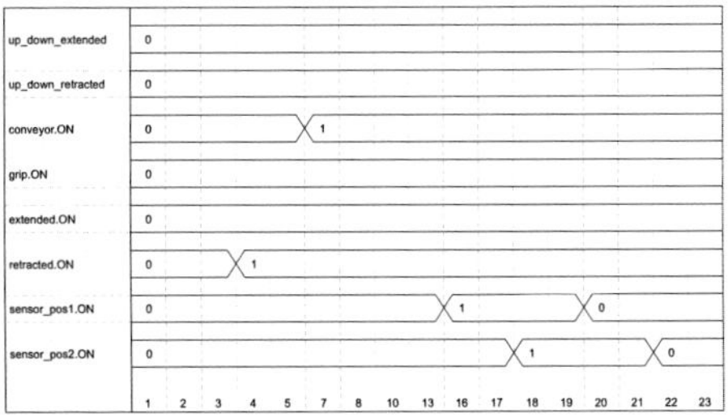

Figure 5.7.: Error path visualization.

more open branches within the dynamic graph. The completeness of the state space can be evaluated at any time. Of course, this assertion is only qualitative because the final size is not certain a priori. For this, no percentage of coverage can be given. Nevertheless, the analysis exposes open trajectories that still have to be considered. Beyond this, dead states are revealed that may result in a deadlock. According to Figure 5.4, a trajectory leading to such a dead state is visualized with an STD ⑤. An example for an error path is shown in Figure 5.7. Note that *conveyor.ON* refers to the plant component and not to the actuator, which switches on the relay of the conveyor drive. The states are a subset of the complete state space of the Gripper Station. In the first four states, the station is initialized. In State 7, the conveyor is started. The pallet reaches *sensor_pos1* in State 16, and the conveyor drive is switched off. Due to the dynamics, the conveyor in the plant possibly does not stop immediately, and the pallet is transported to far. Eventually, both sensors are deactivated in State 23. This state is a deadlock because within the control code, this particular input configuration is not handled. The diagram reveals how this deadlock occurs, and based on it, the control code must be fixed ⑥. Finally, the HiL Simulation has to be started over again until the results are reasonable.
The formal analysis supports the efficient HiL Simulation. However, manual user-interaction is necessary to adapt the control code and to close open branches. Therefore, it can take fairly long to gain a complete coverage of the reachable state space. Nevertheless, a conclusion according to the completeness is drawn and the quality of simulation is evaluated. HiL Simulation is especially suitable if the target controller is available for testing. Otherwise, the SiL Simulation approach is applied as follows.

5.1.2. Software-In-The-Loop Simulation

While performing SiL Simulation, the control software does not run on the target controller but in a simulation environment on a PC. Usually, but not necessarily, this is the same one that executes the plant simulation. It might be applied if the controller hardware is not determined yet. Beyond this, no additional simulation hardware is necessary to interconnect controller and plant. In engineering domains, which apply simulation, like for example automotive or aerospace industry, usually SiL Simulation is executed before HiL Simulation to run software tests. The reason is that MDE approaches are widely established, and for this, it is obvious to test the software model before implementing it on the target controller. However, SiL Simulation does not supersede hardware tests because communication aspects between controller(s) and controlled system(s) cannot be evaluated, in particular.
In general, the SiL framework is equivalent to the HiL approach. The only difference is the execution of the control software within a corresponding engineering environment by simulating the controller hardware. This simulated controller is connected to the simulated plant, and again, the formal model is executed concurrently to capture the dynamic graph. The approach also delivers a qualitative assertion about the coverage of the state space. With this information, the engineer decides whether additional simulation runs have to be performed or whether the control software has to be fixed. Although an assertion of the quality of simulation is possible by the method proposed in this thesis, some weak points remain. The calculation of a complete graph can last fairly long since manual interaction is necessary. Beyond this, controller and formal plant model are not synchronized. If the state space calculation takes longer than the computation of the next controller output configuration, information will be lost. A more critical point is that for HiL as well as for SiL Simulation only the formal model of the plant is considered. The controller in fact executes the control software, but its internal states, which are influenced by flags, counters, or timers, are not considered. Nevertheless, this information is important for a meaningful interpretation of the system's state space. Consequently, simulation remains a tool to test, but it is insufficient to verify. The next section takes up these weak points and presents a more powerful approach to evaluate the correctness of a control software implementation.

5.2. Verification in Closed Loop

Verification is the proof whether a claimed property holds true. According to control systems, verification is the process of proving that the control software meets a given specification and fulfills its intended purpose. Whereas simulation gradually finds its way into the plant engineering domain, there is as good as no application for formal verification. Particularly, it has successfully been used to check military and space applications, but since the verification process still requires a lot of user-interaction, specialist knowledge, and last, not least time, its application has not been justified for manufacturing control systems so far. Beyond this, there is a lack of appropriate tool chains that support the whole verification process, and for this, verification of control systems is still a matter of basic research. However, control software is not just a byproduct anymore but a real cost factor. Constant and high product quality relies on software controlling the plant in an optimal way, and safety can only be guaranteed if the software implementation is correct. Due to reasons of complexity, manual testing methods quickly reach their limits and are not applicable for large-scale systems anymore. However, the wheel should not be reinvented, and because of this, verification should not become an independent step within the workflow of plant engineering, but it should be an integrated part of the whole framework. Therefore, this thesis provides an incorporated approach to perform verification. To do so, the formal models of plant and controller are derived from already existing data as described in Chapter 3. Then, in combination with the formal specification of plant behavior, which is addressed in Chapter 4, model checking is performed as explained in Section 2.7. As it is done for simulation, verification is separated to HiL as well as SiL Verification. Both technologies are explained in this chapter.

5.2.1. Hardware-In-The-Loop Verification

The HiL Simulation procedure, which is described in Section 5.1.1, applies formal technologies to evaluate the quality of simulation. For this, a state space analysis is performed concurrently to the closed-loop simulation. In contrast to it, the state space is not only calculated casually but it is the center point for HiL Verification. To do so, the plant simulation model is replaced by the ${}_{D}$TNCES plant model. The intention of directly interconnecting the formal plant model to the controller hardware is to compute the complete state space of all possible and controllable plant model states and to analyze it according to open trajectories, unwanted states, and the fulfillment of a behavior specification. Doing so, the completeness of the dynamic graph does not depend on simulation runs, but the verification approach itself makes sure to consider all open trajectories by selectively driving the controller through the whole state space. Again, this work is performed by the TMoC. However, in contrast to HiL Simulation, the tool does not only receive the controller output information and calculates the reachable states, but it additionally sends the sensor data of the formal plant model to the controller inputs. Doing so, the closed loop of controller hardware and formal plant model is established.

5.2.1.1. General Remarks

To explain the stepwise state space calculation, the Gripper Station in Figure 3.3 is considered. The pseudocode for the state space computation is given in Algorithm

C.4. As shown in Figure 3.14, the station features four binary actuators and four binary sensors. According to the control software implementation, which is given as an SFC in Figure C.1, the Gripper Station is initialized if the gripper is in the upper position and opened, and if there is no pallet in handling position 1 or 2. For this, the actuator *retract_gripper* is switched on and the sensor *gripper_retracted* is activated. To start the sequence, the controller switches on the actuator *start_conveyor*. The TMoC receives this controller output information and calculates the sub-graph consisting of 10 successor states and shown in Figure 5.8. Thereby, the red-colored states *Z5* to *Z11* are branching states. These states are recognized by comparing the sensor information of each computed state to this one of the initial state of the sub-graph. If at least one sensor value changes, the current state will be labeled as branching state. In addition, the final state of the sub-graph, which is state *Z11*, is also labeled as branching state because it is not necessarily a final state of the whole dynamic graph. In the example, the pallet reaches handling position 1 in state *Z5*, and the sensor *conveyor_pos1* is activated. Moving further, the pallet reaches handling position 2 in state *Z7*, and the sensor *conveyor_pos2* is activated. Then the pallet is transported to the next processing station. For this, first of all the sensor *conveyor_pos1* is deactivated in state *Z9*, and finally, both position sensors are off in state *Z11*.
For each branching state, a path from the initial plant model state to this branching state is derived. The TMoC resets the controller to its initial state for each path and provides the plant output information of each state within this sequence to the controller hardware inputs. If the resulting controller outputs steadily correspond to the expected information, which is given by the states of the sequence, the controller is stepwise driven to the branching state. Then, it provides further output information and the next sub-graph is calculated by the TMoC. Doing so, it is assured that every branching state within the whole system state space is considered. Thereby, loops are recognized and pre- and post-state relations are drawn accordingly.
The HiL Verification framework of this thesis is illustrated in Figure 5.9. Since the analysis is performed by the TMoC, the tool is in the center. The PLC is connected to the PC ① through the SIMBA PNIO device (cf. Section 5.1.1). Formal plant and workpiece models are imported to the tool ②. As stated before, the TMoC reads the controller outputs ① and transfers the information to the formal plant model. Then, it calculates all reachable states based on this initial marking. Afterwards, the plant output information of each identified branching states is transferred back to the controller ①, which evaluates it according to the controller algorithm and provides new output information. This procedure is performed until the state space is complete, that is if there is no branching state without post-state anymore, or if one or more specified final states are reached.
The correct functionality of the control algorithm is evaluated by considering the computed state space. Dead states for example provide hints for possible implementation errors and visualizing the trajectory from the initial to the faulty state reveals the plant

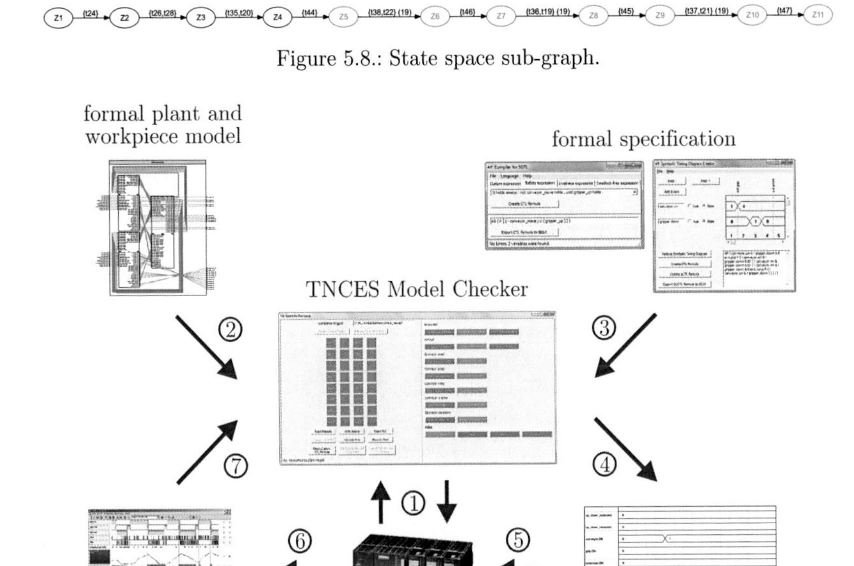

Figure 5.8.: State space sub-graph.

Figure 5.9.: HiL Verification [PLH12].

actions, which have caused it. Consequently, a deadlock can be avoided long before it actually occurs in the real plant. Besides these static considerations, functional and non-functional requirements are checked according to a behavior specification. This is done by formalizing requirements to CTL formulas as described in Chapter 4 and applying them to the computed graph ③. The model checking algorithm returns true if the specification is fulfilled, and a counter example if not ④. This error trajectory is represented in an STD (cf. Section 5.1.1) and shows the path from the initial state to the state, which violates the requirement. With this information, again the control software must be fixed ⑤ and a further HiL Verification iteration is started.

So far, the controller is regarded as a black box system since the HiL Verification approach assumes that controller states are only characterized by the controller output configuration. This assumption is appropriate if the control software does not contain counters, timers, flags or state-dependent variables. However, these elements in fact are contained in the PLC software. To pay attention to this, the corresponding variable information has to be identified by a monitoring software tool ⑥. For the

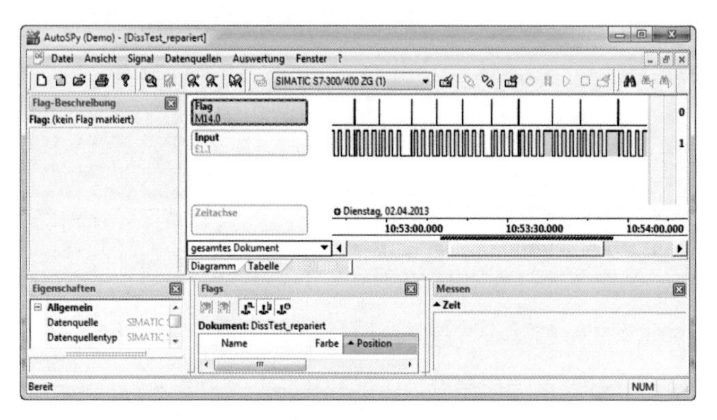

Figure 5.10.: AutoSPy Analysator.

presented framework, the software *AutoSPy*[6] of the GWT-TUD GmbH is applied. The tool adds communication function blocks to the control algorithm and is able to record any controller variable value. Nevertheless, it does not modify the actual control code and deletes the added blocks after finishing the process. Each time a branching state is reached, the TMoC queries the information from AutoSPy and includes it in the state ⑦. Then, after resetting the controller and driving it to the branching state again, not only the controller outputs but also the relevant internal variable values are considered to compare recorded with current state. Selecting a meaningful set of controller variables is the task of the verification engineer. However, the decision process can be supported by a predefined set of crucial variable types. Figure 5.10 shows a screenshot of the AutoSPy tool. It is recording the value patterns of two variable. The flag *M14.0* is used to reset the controller each time the TMoC considers another branching state. The input *E1.1* is referred to a sensor of the Jack Station.

5.2.1.2. Case Study

The practicability of the HiL Verification framework is investigated in a case study. The tests are performed on different computers to compare the results and especially to consider the influence of computation power. For this, the four different configurations, which are summarized in Table 5.2, are taken into account. As a first result, it turns out that the performance of the HiL framework is hardly influenced by PC's CPU clock speed because latency is particularly impacted by timing constraints within the control code as well as by the signal propagation time on the communication channel. Of course, this does not hold for SiL Tests, and for this, calculation time is reduced by

[6]AutoSPy website. URL, http://www.autospy.de/en/index.html, November 2013.

#	operating system	CPU	RAM
1	Windows XP 32 bit	Intel® T2500 @ 2.0 GHz	2 GB
2	Windows 7 32 bit	Intel® Core™ 2 Duo P8400 @ 2.26 GHz	3 GB
3	Windows 7 64 bit	Intel® Core™ i5-M520 @ 2.4 GHz	4 GB
4	Windows 7 64 bit	Intel® Core™ i5-2300 @ 2.8 GHz	8 GB

Table 5.2.: PC systems for case study.

#	model description	transitions	places (total)	places (timed)
1	Gripper Station (not initialized)	57	63	7
2	Jack Station + Slide Station (not initialized)	98	93	9
3	Gripper Station + Jack Station + Slide Station (not initialized)	155	156	16
4	Gripper Station + Jack Station + Slide Station (initialized)	155	156	16
5	Gripper Station + Jack Station + Slide Station + Store Station (initialized)	166	168	19
6	complete demonstrator (initialized)	332	336	38

Table 5.3.: Models for case study.

magnitudes. Nevertheless, HiL Tests are inevitable to evaluate the actual implementation of the controller, not least because engineers are practitioners and demand a comprehensible proof of correctness regarding the final product. For this, considering the real controller's correct implementation is a crucial point for the framework to be applied in practice.

To quantify performance and to study the impact of computation power, model size, and timings, the different substations of the demonstrator (see Section 3.1) are first considered separately and then in combination. In principle, the experiments also can be projected onto real-scale plants because simulation hardware and simulation as well as verification software are able to handle an arbitrarily - and practically meaningful - quantity of signals. The only constrains are computer resources and computation time. Admittedly, both shall not be underestimated. Table 5.3 gives a survey of the six different plant models, which are analyzed. The control algorithm of the stand-alone Gripper Station (1) is depicted in Figure C.1.

The state space computation results are summarized in Table 5.4. The different computation times are assigned to the computer systems in Table 5.2. As noticeable, the calculation times for the different models do not significantly depend on CPU speed. Of course, this fact was to be expected because the resources of a computer are much higher than of a PLC. Since the verification tool is written as a single core application, advanced CPUs would surely deliver similar results. The real bottleneck is

model	states	computation time	
1	663	PC system (1): 1 min 3 s	PC system (2): 1 min 14 s
		PC system (3): 1 min 13 s	PC system (4): 1 min 1 s
2	1135	PC system (1): 7 min 18 s	PC system (2): 8 min 37 s
		PC system (3): 8 min 32 s	PC system (4): 6 min 35 s
3	4460	PC system (1): 18 min 15 s	PC system (2): 20 min 49 s
		PC system (3): 21 min 28 s	PC system (4): 16 min 3 s
4	1820	PC system (1): 12 min 55 s	PC system (2): 15 min 34 s
		PC system (3): 15 min 24 s	PC system (4): 11 min 58 s
5	1566	PC system (1): 11 min 21 s	PC system (2): 13 min 33 s
		PC system (3): 13 min 15 s	PC system (4): 10 min 27 s
6	10115	PC system (1): 128 min 29 s	PC system (2): 150 min 20 s
		PC system (3): 147 min 38 s	PC system (4): 127 min 6 s

Table 5.4.: HiL Verification results.

the Ethernet communication channel between SIMBA PNIO device and verification PC. The tests on all computers reveal that the minimal time resolution t_{input} is about 70 ms, which is the time necessary to recognize a change of a controller output. Consequently, this is also the shortest possible delay for timers within the control code, so that the TMoC is able to distinguish the two states before and after the timer is expiring. Lower values led to loss of information while performing the experiments. Nevertheless, this lower limit should be fast enough for most applications in manufacturing industry. If not, another communication protocol will have to be applied.
Whereas the communication delay is technically caused, a further time delay is explicitly related to the control code. The TMoC provides input information from the plant model to the controller. Then, it polls the controller outputs and recognizes changed values. Because of the controller being a black box while performing HiL Verification, the TMoC has no clue after which time it shall stop polling. This appears especially if delays are included in the control code, but also if the cycle time is relatively high. For this, the TMoC has to wait at least until no state change is realizable anymore. The problem becomes visible regarding the Gripper Station. For example, the gripper is moved down and shut to grab a tin. To make sure that the workpiece is gripped properly and to pay attention to the plant dynamics, the time delay t_{delay} of 70 ms is implemented in the control code. Afterwards, the gripper is moved up. For the tool, this means that it has to wait for 70 ms to recognize the first state change (Step6 to Step7 in Figure C.1), 70 ms for the timer (T8 in Figure C.1), and another 70 ms to recognize the second state change (Step7 to Step8 in Figure C.1). Accumulating the times, one receives a delay t_{PLC} of about 210 ms. That means, each time the controller outputs are polled, the verification tool has to wait as long before continuing. Hence, the longest determined time delay is also this one for the overall HiL Verification pro-

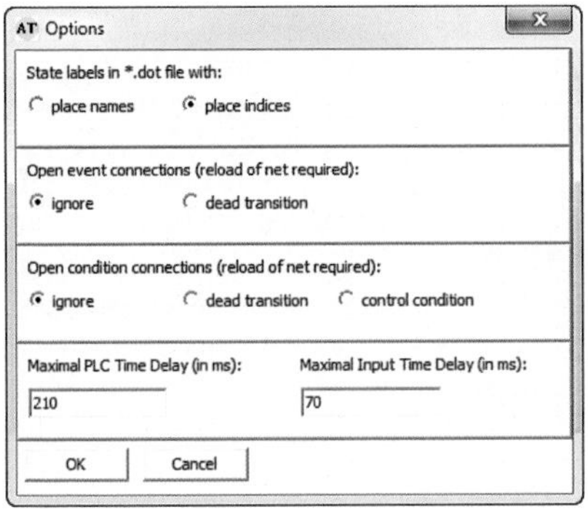

Figure 5.11.: Options dialog of the TMoC.

cedure. As shown in Figure 5.11, both time delays are considered in an options dialog. Thereby, the *Maximal Input Time Delay* is the time delay t_{input} on the communication channel. The *Maximal PLC Time Delay* is the time frame t_{PLC} for cyclically polling the controller outputs. For this, it is determined as $t_{PLC} = 2\ t_{input} + t_{delay}$.
In practice, t_{PLC} elapses each time the TMoC provides new controller input information and polls the resulting controller outputs. As visualized by red-colored states in Figure 5.8, this is always necessary if a plant output, or a sensor respectively, changes its state. To optimize this task, the structure of the sub-graph is utilized. As shown in Figure 5.12, each sub-graph consists of one starting state z_a and one or more branching states z_b. The starting state z_c of the next sub-graph is equal to z_b except that the plant input information is overwritten with the polled controller output information. Thereby, all successor states of z_c represent the plant reaction to this particular controller output configuration. The located branching states z_{d1} and z_{d2} contain the new plant output information, which has to be provided to the controller inputs. However, all states of sub-graph 2 have got the same initial state z_c. Consequently, the data basis is consistent, and for this, not every branching state has to be considered separately. These ones, which contain the same sensor information, are combined to a set. For the example of Figure 5.12, let z_{d1} and z_{d2} be branching states with different plant component states but with the same sensor state. The TMoC provides this plant output data to the controller only once and receives new controller output information. Then, z_{e1} and z_{e2} are derived by copying z_{d1} and z_{d2} and overwriting the plant model inputs with the polled controller output information. Thereby, no information is lost, but the procedure is performed only once for each calculated sensor configuration set of one sub-graph. Doing so, the number of TMoC and PLC interactions is reduced from seven to two for the example sub-graph given in Figure 5.8. This saves approximately 1 s of "waiting" time plus the time, which is necessary to drive the controller

to the branching state. Consequently, the execution time for the HiL Verification is reduced significantly due to the number of computed sub-graphs.
The time to consider branching states actually has the biggest impact on calculation time. It elapses while resetting the PLC and driving it to the branching state each time such a trajectory is regarded. Therefore, the time for analysis grows the bigger a plant model gets and the more complex the control software becomes. Due to the HiL Verification algorithm, this is the Achilles heel of the approach and limits the plant size, which can be analyzed in a meaningful time. Nevertheless, the plant demonstration example shows that a system of such size is easily manageable.
The state space calculation has to be performed only once for one model. The subsequent application of the formal behavior specification to it is carried out independently. Beforehand, the state space delivers information, which shall be considered. In ideal case, the verification tool returns a complete graph, which either has no open trajectories, or which ends in one or more specified final states. The models (1) to (6) all end in their last state, which means that the production sequence has been carried out successfully. However, there are some additional dead states within the graph, which have to be analyzed. To support this work, each error paths is visualized in an STD, which is described in Section 5.1.1, to reconstruct how the errors have occurred. It turns out that all produced dead states are false negatives. That means that they indeed appear in the model but have no practical relevance. The cause is as follows: The PLC switches the conveyor on. If a pallet reaches a position sensor, the controller will turn the conveyor off. Because of the modeled dynamics within the formal model, it is possible that the conveyor moves on and stops the pallet in an unwanted position (cf. Figure 5.7). Technically, this problem does not have an impact on the plant operation because of the fairly slow conveyor speed. Nevertheless, this could be a technical issue in another plant, and for this, the state space analysis gives first hints about critical plant behavior.
To complete the analysis, finally the specified requirements are applied to the computed state space. Exemplary, the temporal logic formulas 4.1, 4.2, 4.3, and 4.7 from Chapter 4 are considered. As shown in Figure 5.13, the formulas are entered in a dialog, either manually or per copy and paste from the specification tools. Several formulas can be entered separated by a blank line. The TMoC analyzes each formula and gives feedback in the dialog window whether the formula is fulfilled or not. The computation time highly depends on the complexity of the nested formulas and on the number of states, which are taken into account. For this, it is hard to make an absolute assertion about calculation costs. Table 5.5 summarizes the results of the HiL Verification case study. Since all formulas are related to the Gripper Station, model (2) is not considered. The test results are depicted for PC system (4) only. Like for the state space calculation, the other systems show similar results according to computation time.
The production requirement, which is specified in Formula 4.7, holds true for each model. This was to be expected as the state space calculation always terminated in

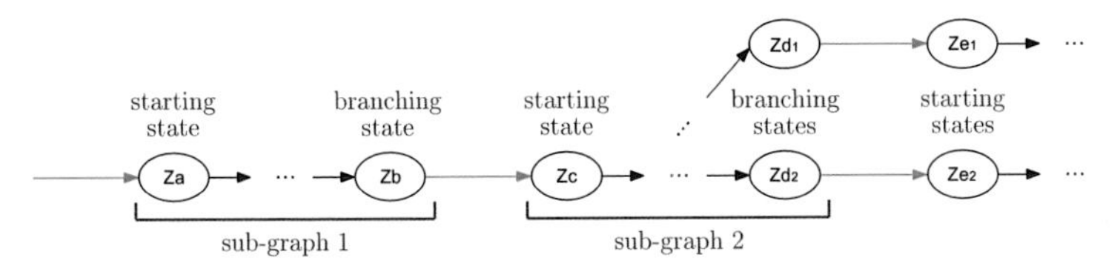

Figure 5.12.: Sub-graph composition.

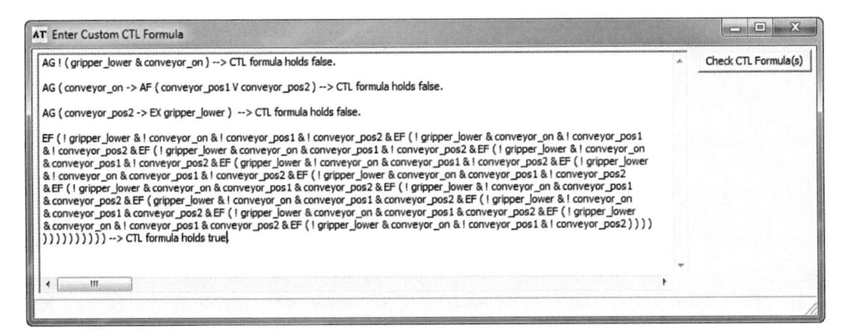

Figure 5.13.: Specification dialog.

one specified final state. In contrast, Formula 4.1, 4.2, and 4.3 are evaluated to false. The safety requirement of Formula 4.1 is violated because of the control algorithm implementation depicted in Figure C.1. The conveyor is moving in state *Step5*. After reaching handling position one, it stops and the gripper is immediately extended in state *Step6*. Due to the modeled dynamics of the conveyor, it does not stop instantly. For this, there is actually a state, in which the conveyor is moving and the gripper is extending. Nevertheless, this is not critical in the real plant because it has been considered during positioning the sensors. Consequently, it is sufficient for the plant to fulfill the requirement that the gripper is not extended while the conveyor is moving. The corresponding property is given in Formula 5.1. This adapted safety requirement holds true for all five models.

$$\textbf{CTL: } AG\ !\ (\ \text{gripper.up\_down.Extended}\ \ \&\ \text{conveyor\_on}) \tag{5.1}$$

Formula 4.2 is violated because there exists a state, in which the conveyor does not stop after the pallet has reached a position sensor. As already discussed, this is a false negative because it has no practical relevance. Nevertheless, the model checker only returns true or false, and it is not visible whether there are true negatives. To receive this information, there are two possibilities to proceed. On the one hand, the model can be refined and analyzed again. However, this approach needs reengineering and

model	**computation time and results**			
	Formula 4.1	Formula 4.2	Formula 4.3	Formula 4.7
1	< 1 s (false)	< 1 s (false)	< 1 s (false)	< 1 s (true)
3	< 1 s (false)	13 s (false)	2 s (false)	13 s (true)
4	< 1 s (false)	< 1 s (false)	< 1 s (false)	4 s (true)
5	< 1 s (false)	< 1 s (false)	< 1 s (false)	2 s (true)
6	5 s (false)	59 s (false)	5 s (false)	61 s (true)

Table 5.5.: HiL benchmarks results of application of specification.

naturally produces a different state space, which is hardly comparable to the one of the original model. On the other hand, the state space graph can be modified directly. The idea is to exclude the false negatives, which means to eliminate the unwanted dead states plus the corresponding trajectories, and apply the specification to this adapted state space graph. The implementation is straightforward since the pre- and post-state relations have to be deleted only. This is done by considering a particular dead state and going backwards in the state space graph. Thereby, each state along the trajectory, which has only one post-state - that is the dead state -, is deleted as well until a state with at least one additional post-state is found. Finally, the relation of this live state to the dead state trajectory is deleted. For this, the branch in the state space, which leads to the dead state, is fully eliminated. This is automatically done for each false negative by identifying the particular state. Afterwards, Formula 4.2 is applied to the adapted state spaces of all models and returns just one false positive for each graph. This is produced after finishing the gripper sequence. The conveyor is switched on and the pallet heads on to the next station. In this case, both position sensors of the Gripper Station are in state off. According to the model and to the plant, this behavior was to be expected.
Regarding Formula 4.3, the modification of the state space is no solution because the formula is too strict. Within the model, there is no state transition from place *conveyor_pos2* labeled to place *gripper_lower* labeled. Because of this, the formula is rewritten to Formula 5.2 to directly consider the state transition from a triggered sensor to an activated actuator. Thereby, $E0.2$ and $A0.0$ are the plant model outputs and inputs as depicted in Figure 3.14. The analysis of the formula delivers the same false positive as Formula 4.2, which was to be expected as well.

$$\textbf{CTL: } AG\ (\mathrm{E0.2} \rightarrow EX\ \mathrm{A0.0}) \tag{5.2}$$

After having calculated the state space of the closed loop and having applied the formal specification to it, the controller hardware is attached to the demonstrator. Regarding Figure 5.1, this is done by disconnecting the purple Profibus cable from the SIMBA device ⑥ and connecting it to the ET200M I/O station ②, which is connected to the real plant sensors and actuators. Since all specified properties are

fulfilled and the state space calculation ends in its specified final state, the control software is analytically proven to be correct. Consequently, after attaching it to the real plant, it behaves as expected and no further code changes are necessary.
The results of the HiL Verification case study approve these ones of the HiL Simulation. Due to the computation of the whole state space, critical scenarios are revealed in addition. Although they are not relevant for the demonstration example because of the relatively slow plant parts' movement, they could be of interest for other manufacturing systems. Since the verification procedure runs fully automatically, the state space is produced without user-interaction. The engineer just has to provide a meaningful specification and has to evaluate the results of the formal analysis. However, involving the internal variables of the controller is a challenge because additional tools have to be applied to record and to consider them. Especially, this arises according to timers because tolerance ranges have to be specified to compare the captured and the current data. This task is time consuming and costly. For this, a further verification approach is introduced in the following.

5.2.2. Software-In-The-Loop Verification

Due to the complexity of plant engineering, it is appropriate to divide tasks and to parallelize their treatment. In practice, this procedure is well applied, and for this, software engineering environments support software implementation and testing without actually being dependent from the target hardware. Naturally, the HiL Verification approach is not applicable in this case. Therefore, SiL Verification provides a promising tool. To do so, the controller is not considered as a black box, but its behavior is formally modeled and therefore fully observable.

5.2.2.1. General Remarks

Figure 5.14 illustrates the SiL Verification framework of this work. The first step is the formalization of control code ① (see Section 3.3). Afterwards, the derived controller model is composed with the formal plant and workpiece models to a closed-loop system model ② by interconnecting the particular condition inputs and output as shown in Section 3.4. The subsequent state space computation is performed by the TMoC. This task is performed automatically and does not require user-interaction. Having calculated the entire state space, the approach is in fact similar to HiL Verification. Besides the static analysis of deadlocks, properties in terms of a formal behavior specification are checked. For this, the specification is applied to the computed graph ③. The model checking algorithm returns true if all properties hold true, and false if at least one is violated. In the latter case, a counter example is produced and visualized in an STD ④ to show the trajectory from the initial to the error state. Finally, the control code must be fixed with this information ⑤, and the process starts over again.

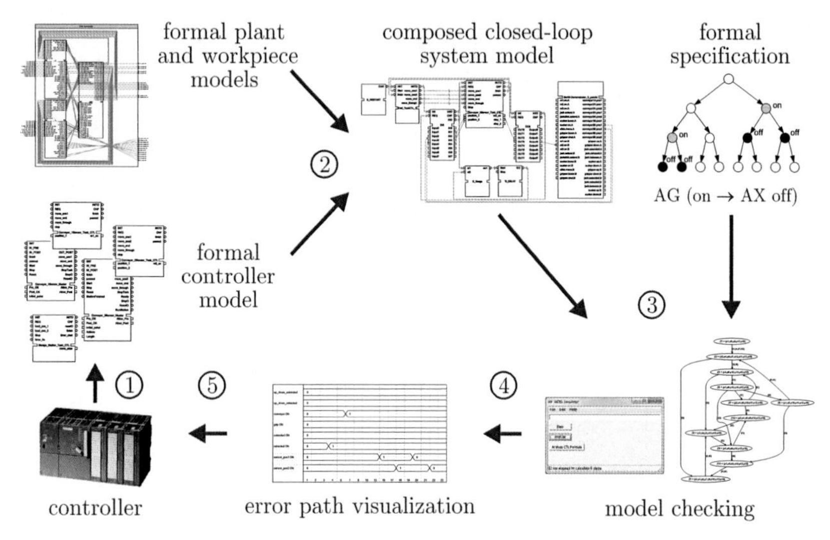

Figure 5.14.: SiL Verification.

SiL Verification has an essential benefit over the HiL approach because no additional hardware is necessary to interconnect controller and plant model. Furthermore, the controller states are fully observable since the controller is translated to a formal model as well. For this, internal variables, flags, and timers are contained in the model, and it is not necessary to read out this information with additional software tools. While performing HiL Verification, the controller outputs are continuously polled by the verification tool. In this connection, a crucial point is the upper time limit to decide whether the controller is still processing, due to timers, counters or other dynamic function blocks, or is trapped in a livelock. Consequently, the upper time limit has to be specified for each control program to get meaningful verification results. Being based on a different approach, this problem is insignificant for SiL Verification. Because of this and since the CPU of a desktop PC is much more powerful, one can expect a faster state space generation in practice. Nevertheless, the SiL approach indeed considers the functional behavior of the system under control but not its dynamics according to cycle time or communication duration. For this, SiL Verification does not supersede hardware tests because these parameters are crucial for practical evaluation.

#	model	transi-tions	places (total)	places (timed)	states	computation time
1	Jack Station	175	146	16	9348	3 min 56 s
2	Gripper Station	119	113	11	2301	32 s
3	Store Station	52	41	4	402	7 s
4	Gripper + Jack + Store Station	346	300	31	16729	8 min 58 s
5	complete demonstrator	692	600	62	34914	39 min 33 s

Table 5.6.: SiL Verification results.

5.2.2.2. Case Study

The practicability of the SiL Verification framework is investigated in a case study. The time, which is necessary for computation, is only determined for computer system (4) of Table 5.2. Since the TMoC is a single core application, the other systems would deliver similar results according to calculation time. Table 5.6 summarizes the design parameters of the different models and the results of the state space computation. The procedure is performed for the stand-alone Jack (1), Gripper (2), and Store Stations (3), as well as for the combination of the three stations (4). Finally, the whole demonstrator is considered (5). The results prove that the closed loop of plant and controller models is analyzable in an appropriate time. Admittedly, the TMoC is not focused on an optimal usage of computer resources. Rather, it supports the model checking framework and includes all the aspects of this thesis' approaches. Consequently, the computation time is surely reducible by paying more attention to an efficient code implementation. However, this is a point for future work.
Whereas the applied modeling approach is just a means for the HiL Verification, it reveals its full potential while performing SiL Verification. First of all, modeling of discrete time intervals keeps the model clearly arranged. This point is discussed in Section 3.2 for the cylinder module in Figure 3.4 as well as for the conveyor belt module in Figure 3.9. The dynamics of the plant components is represented in minimal models according to the number of places and transitions. To do so, time intervals are added, which implicitly represent intermediate states. Of course, the movement of a plant component is not discrete, and in principle, an unlimited amount of states could be used to describe it within the formal model. Nevertheless, there is little information inside such a detailed model because for the controller only sensor values are of importance. On the other hand, considering just the different sensor states as it is done for the conveyor belt in Figure 3.9 would not be appropriate without temporal information. In the real system, the controller must be faster than the controlled plant to recognize every state transition. However, the conveyor belt module would need

model	**computation time and results**				
	Formula 4.1	Formula 4.2	Formula 4.3	Formula 4.7	Formula 5.1
2	< 1 s (true)	5 s (false)	< 1 s (false)	34 s (true)	< 1 s (true)
4	3 s (true)	21 s (false)	3 s (false)	1 min 23 s (true)	3 s (true)
5	6 s (true)	6 min 8 s (false)	7 s (false)	11 min 7 s (true)	6 s (true)

Table 5.7.: SiL benchmarks results of application of specification.

only four steps to assume each possible state. Because of the comparably-complex controller model, a pallet on the conveyor would have left it long before the controller was able to notice even one sensor change. Because of this and to project the demand of a controller, which is faster than the plant, the discrete intervals are added to model the implicitly-intermediate states. In addition to this dynamic consideration, the closed loop of formal plant and controller models contains many concurrently-running processes. Determining the firing of all possible transitions would consequently blow up the dynamic graph. Nevertheless, there is no absolute simultaneousness in a real system, and for this, such a view on it is unnecessarily complicated. By adding time intervals to the model, concurrency is reduced. For this, many parallel trajectories in the state space graph, which contain the same set of transitions but in sequences that are mutually distinct, are combined. Of course, these intervals have to be applied in a meaningful context, but they help to cope with the state space explosion problem.
Another benefit is the consideration of event signals to synchronize the firing of transitions. Without such a capability, analyzing an IEC 61131 control device would quickly produce a huge set of reachable states because of the cyclic execution behavior. Thereby, the actual information is not increased because updated sensor as well as internal variable values are only recognized in a small number of executed cycles. To factor out analytically-useless states, event signals are included in the formal controller model to recognize sensor information updates and internal state transitions (see Figure 3.24). Thereby, the execution of the control functions is sequentially performed by firing events (see Figure 3.25) to narrow the number of concurrently-executable steps. Furthermore, event signals are suitable to model mechanical interconnections in the plant model. Doing so, for example an end position sensor is not activated in a successive state after the plant component has reached the corresponding end position, but both state transitions are synchronized and considered in one step. Of course, this holds for modeling the workpiece behavior as well since plant components and workpieces are connected through event signals as described in Section 3.2.1. Doing so, again attention is paid to the physical interconnection of components within the real plant.
The results of the state space calculation prove that ${}_{D}$TNCES are an appropriate tool to model a manufacturing system. Contrary to the expectations that time-evaluation

complicates the analysis, it turns out that the state space is decreased by applying meaningful intervals. In addition, the model is kept well arranged since the dynamics is represented implicitly. Beyond these semantic advantages, also the component-based and hierarchical character of $_D$TNCES serves well for model generation from the engineering point of view, which is discussed in Chapter 3. The state space analysis reveals the great benefit of the SiL approach. Since the controller is not a black box but its behavior is fully represented by the formal model, of course no additional analyzing software is necessary. Furthermore, the false negatives, which are related to the conveyor belt in the HiL Verification procedure, do not appear performing SiL Verification. For this, the controller model recognizes every value change of the position sensors and reacts to it before the pallet is transported too far and a false negative could be produced. This corresponds to the behavior of the real system and is achieved by modeling the dynamics with discrete time intervals. The computation of the state space of the whole demonstrator reveals only one dead state. This is after the pallet has left the conveyor of the second Store Station. As this is the predefined final state, the calculation is performed successfully.
Finally, the specification of behavior is applied to the state space of the closed-loop model. As it was to be expected, model checking delivers similar results like for the HiL Verification. Due to the bigger model size and the resulting increased dynamic graph, the analysis of a CTL formula last longer if performing SiL Verification. However, the specification is still checked in a meaningful time as it is visible in Table 5.7. The safety requirements of Formulas 4.1 and 5.1 are both evaluated to true. Since the false negatives of the HiL Verification do not occur in the state space, which is produced by SiL Verification, the liveliness requirement of Formula 4.2 returns only one counter example. This is, after the pallet has left handling position 2 of the Gripper Station and is forwarded to the Store Station. This behavior was to be expected. However, to have a meaningful liveliness requirement, Formula 5.3 is checked in addition. It expresses that if the gripper is moved down, it will either be closed or opened in a future state. For this, it is excluded that the gripper moves down and remains in this position without actually processing the workpiece. This requirement is evaluated to true. Because of the higher complexity, the evaluation time is increased to about 12 minutes for the whole demonstrator.

$$\begin{aligned}\textbf{CTL:}\ & AG\ (\ \text{gripper.up\_down.Extended}\ \rightarrow AF\ (\ (\ \text{gripper.grip.Retracted}\ \&\\ & AF\ (\ \text{gripper.grip.Extended}\ )\ )\vee(\ \text{gripper.grip.Extended}\ \&\\ & AF\ (\ \text{gripper.grip.Retracted}\ )\ )\ )\ )\end{aligned} \tag{5.3}$$

The deadlock-free requirements of Formulas 4.3 and 5.2 are both evaluated to false. The reason again is that the *EX* operator is too strict. Because of this, it is rather meaningful to formulate such a requirement as a liveliness one by replacing *EX* with *AF*. Doing this for Formula 4.3, the property is evaluated to true in about 6 minutes for the whole demonstrator. Last, the production sequence of Formula 4.7 is applied

to the model. This functional requirement is evaluated to true as it was to be expected.
The results of the SiL Verification case study confirm these ones of the HiL Verification. Thereby, the specific advantages of applying the formal closed-loop system model become apparent as discussed in this section. However, the most crucial ones are a faster generation of the state space and the independence from the controller hardware. The SiL Verification approach especially supports software engineering in a very early project phase, in which the target controller hardware is often not available or not yet determined. Beyond this, it is applicable to migrate existing controller structures by verifying the control software without being in need for the actual controller hardware. Again, formalisms and dull theory are embedded into the TMoC to arrange the procedure as user-friendly as possible. Like for HiL Verification, it also holds for SiL Verification that if correctly applied, all properties that are verified for the closed-loop model will hold for the real controlled process as well.
Nevertheless, formal verification features limits of application. So far, it is a scientific research field, and although, many approaches have been established over years, they are not yet applied in the plant engineering domain. Therefore, the subsequent section discusses reasons for this circumstance.

5.2.3. Limits of Verification

Studying verification and model checking approaches, the reader, who is in step with current practice, might wonder why they are not already widely applied in daily engineering, although they promise theoretically optimal solutions. To discuss this question, this section shall point out some issues concerning the limits of verification.
The most obvious one from the analytical point of view is probably the state space explosion problem. Since every possible system state has to be considered, the state space rises exponentially according to system size. This makes explicit calculation very inefficient or even impossible for large-scale systems. The complexity can be influenced by effective modeling strategies. For example, considering the closed loop of controller and plant limits the state space because the controller is not verified for any possible input configuration in any state but just for physically-possible and meaningful input information provided by the plant model outputs. Of course, this adds further state information of the plant model, but nevertheless, the complete state space is limited in practice. Furthermore, the structure of the models is crucial for efficient verification. Since $_D$TNCES are component-based, hierarchical level-by-level verification as described in [MHH07] is possible. The approach presented in [WLH11] also benefits from the $_D$TNCES structure and narrows the state space by searching for transition invariants and deriving a reduced representation of the net. Usually, this *transition-invariants graph* (TIG) is smaller than the explicit state space graph and easier analyzable because of its reduced size.

Due to the system's complexity, verification is performed by computers. However, proving implemented algorithms with implemented algorithms seems to be conflicting if the correctness of the verifying one is questionable. Consequently, to be usable for industrial applications, there is a need for software certification. So far, there is no standard, which prescribes the application of verification. Therefore, the verification engineer has to be familiar with the particularly-applied approach.
Regarding the inputs for verification, a correct model is absolutely necessary because all assertions are made referring to it. Since a model (yet) cannot be automatically abstracted from the real system, the plant engineer has to pay very much attention to derive a meaningful one. Furthermore, the specification has to contain all relevant test cases and properties to be checked. Otherwise, the verification results might be considered as insufficient.
In spite of these weak points, it is the opinion of the author that formal verification may fruitfully extend plant engineering according to the overall design and engineering process in general and to the control software correctness in particular. Nevertheless, it does not factor out the engineer who has to critically evaluate the results of verification. To sum up this chapter, the next section draws a conclusion.

5.3. Summary

This chapter combines all the approaches of this thesis to study the behavior of the closed-loop system. Thereby, the two analyzing technologies of simulation and verification are considered.
Simulation is a powerful tool to test a control software implementation. However, conventional simulation is not sufficient to evaluate the quality of the simulation results since there is no indicator according to completeness. Because of this, an advanced approach is proposed in this chapter to extend HiL as well as SiL Simulation. The idea is to additionally apply a formal model, to connect it to the controller, and to calculate the state space of the system. The resulting graph shows open trajectories that further have to be analyzed and reveals dead states, which might belong to a deadlock. With this information, the software must be fixed or additional simulation runs have to be executed. To automate this work, the software tool TMoC is implemented in the scope of this thesis. It computes and displays the complete reachable state space of the interconnected plant model and controller hardware periphery. Thereby, it keeps theoretical aspects in the background, and because of this, it is capable of being integrated to the plant engineering workflow.
Nevertheless, achieving a complete dynamic graph can become a rather time-consuming task. Therefore, this manual user-interaction is completely carried out by the TMoC while performing verification. Again, the two approaches of closed-loop HiL and SiL Verification are considered. Both have specific advantages and consequently are tai-

lored to different fields of application. The HiL Verification is used to analyze the controller hardware under conditions, which are very close to reality, whereas the SiL Verification considers all possible states of the closed loop of plant *and* controller models. The application of the two approaches is investigated in case studies. The results of the experiments prove that verification is embeddable into the workflow of daily plant engineering by an integrated usage of available data.

A further concern of this chapter is to show how to interpret the formal results of analysis, and particularly, how to present them in an understandable way. On the one hand, the state space graph of the closed-loop model is visualized to ease the evaluation of completeness. On the other hand, the error path of a violated property is graphically treated to enable reconstructing the cause of failure. This information is crucial for a fast and systematic error diagnostics.

After having verified the control software, the controller hardware has been attached to the real demonstrator. The controller behaved correctly without requiring any additional code changes. Concluding the results of the case studies, the HiL Verification approach is the more practical one. This is because the controller hardware and consequently the actual software implementation are considered explicitly. For the formal analysis, this reveals a further advantage. The plant model is exactly the same for the HiL and for the SiL Verification. However, the controller only is represented through its in- and outputs for the HiL approach. Therefore, the dynamic graph consists of the plant model states in combination with the controller in- and output states. Assuming that controller hardware and controller model have the same execution behavior, the HiL dynamic graph is a sub-graph of the SiL one because the concurrent behavior of the formal controller model is figured out. Usually, this results in a smaller dynamic graph for the HiL Verification approach, and consequently, HiL Verification can handle much bigger systems than SiL Verification.

Although, the control code does not have to be formalized for the HiL approach, the internal states of it could be important for the verification process. Because of this, an additional analyzing tool must be applied. Of course, this does not hold for SiL Verification because the closed loop of plant and controller models is taken into account. Beyond this, SiL Verification does not require additional hardware to interconnect controller hardware and verification PC. For this, it is the opinion of the author that both approaches have their specific advantages and should be applied according to the general requirements. Therefore, the HiL Verification will be used if a plant is designed from scratch and the controller hardware is available for testing. The other way round, SiL Verification will be deployed for migrating and modernizing existing plants. Here, the controller is not accessible for testing, and for this, a model of it is considered. Nevertheless, the successfully-performed verification depends on meaningful models and on a formal specification, which considers all critical scenarios. Consequently, it remains a tool to analyze system's behavior, but it does not supersede the work of engineers.

6. Conclusion and Outlook

This chapter summarizes the research results of this thesis. First of all, the findings are concluded, and subsequently, some subjects of future work, which can be performed based on this work, are provided.

6.1. Conclusion

The main objective of this thesis has been to establish a framework, which includes formal analysis to the workaday life of plant engineering. Thereby, the most important contributions are:

- A guideline how to implement plant and controller models with the formal modeling language of $_D$TNCES. Beyond this, deriving formal models out of informal data is achieved through utilizing existing data and treating it for semi-automatic model generation. The formal plant model generation is implemented in the software tool IED Converter. For the controller model generation, the principle is described.
- Two specification languages - a graphical and a text-based one - for the formal description of plant behavior. Both are translated to temporal logic formulas to be applied for the model checking process. To support the specification, the approaches are implemented in two software tools, namely the Compiler for SOTL and the STD Editor.
- An extension of conventional closed-loop plant simulation by evaluating the completeness of simulation. To do so, a reachability analysis is performed concurrently to the simulation process. The recorded dynamic graph delivers indications whether the simulation covers the whole possible system behavior or if there are open trajectories, which still have to be checked. This analysis is implemented in the software tool TMoC.
- A model checking framework to verify the correct implementation of the control software. It combines the approaches of this thesis to an integrated formal analysis, which is implemented in the software tool TMoC as well.
- A proof of concept by performing a case study on a technical plant in lab-scale.

The research work shows that the combination of simulation and verification is absolutely meaningful. Simulation provides a visualization of the physical system model and facilitates testing whether the plant construction is realizable. The execution of concrete test cases allows determining whether a specified production scenario can be carried out. Thereby, the extension of conventional simulation by formal analysis enables evaluating the quality of simulation. However, especially the consideration of open trajectories can take fairly long. For this, verification is the more powerful tool to evaluate the correctness of the control software because it handles this task automatically.

Comparing the HiL and the SiL approach, the HiL is the more extensive one due to the mandatory interconnection of controller hardware and plant simulation. This is because additional hardware in terms of the SIMBA PNIO box is required. Seemingly, the SiL approach is the more favorable one if the additional costs for controller modeling are factored out. Actually, SiL tests are applied before HiL tests in industrial practice, but however, hardware tests are crucial for safety considerations as well. The case studies of this thesis show that the interconnection of the discrete plant and controller models is indeed sufficient to evaluate the logical correctness of the control code, but assertions according to temporal constraints can hardly be done. Whereas plant dynamics and the delay of controller functions are modeled with discrete time intervals of $_{D}$TNCES, the communication between controller and plant is not regarded at all. Nevertheless, the HiL case study reveals that these occurring delays are not negligible. As these impacts only can be analyzed by considering the concrete hardware, it is the opinion of the author that HiL and SiL tests complement one another in a meaningful scope.

The framework is tailored to the demands of the engineering practice. It does not overburden with dull theory, but it encapsulates formalisms into domain-specific description tools, which engineers are used to. In addition, it does not open a completely new branch to analyze the control software, but it is integrated to the already-established workflow. Thereby, a crucial point is that it does not change the actual control code, but it reveals failures by evaluating the closed-loop system behavior. The engineer keeps control of the overall process and performs changes manually. This corresponds to the general distrust of engineers against automatic code changes, and thereby, the acceptance is generally increased.

Plant engineering is performed in team work between engineers of different technical domains. The framework pays attention to this circumstance and supports assembling the different work packages. For example, the specification is written by one team member, whereas the plant model is developed by another one. A third one prepares the control code and at the end of the day, all parts are joined. Thereby, errors of one team member do not propagate and are revealed at an early stage. As the research work for the verification of manufacturing control systems is still ongoing, subjects for future work are discussed in the next section.

6.2. Outlook

The establishment of an analyzing framework is evolutional. Concepts can be substituted by more efficient ones, and formalisms can be extended with novel findings. In addition, the applied tools can evolve and their design can be improved. Therefore, this thesis' framework is modularly arranged so that adaptation is supported and even requested. To list some open issues, the following survey focuses on extensions, which can enhance the framework prospectively.
The first point is related to the automatic model generation. On the one hand, the modeling and simulation tool vendors should improve the capabilities to export and import CAD data by applying a standardized exchange format. On the other hand, the direct translation of a CAD model to a formal $_D$TNCES could be implemented as well. The formal controller model generation from a concrete PLC implementation is another open issue. The translation rules have to be implemented in a compiler to derive a model from the control code automatically. Both improvements would accelerate the modeling process and delimit the need for user-interaction.
Model checking only delivers meaningful results if the specification of the properties is sufficient. Fulfilling all specified requirements does not necessarily mean that the system works correctly. There could be properties, which have not been considered but which are mandatory to be proven. To evaluate the completeness of a specification - that is its coverage - several approaches exist. Benchmarking and including them to the framework would extend the expressiveness of the verification results.
The quality of simulation is rated by concurrently computing the state space of the closed-loop system and evaluating it according completeness. As it has been discussed, considering all possible trajectories can last fairly long because additional simulation runs have to be carried out manually. Therefore, an automatic test case generation would support this task. To do so, the dynamic graph would be the basis for deriving the trajectories, which still have to be checked. The TMoC would drive the controller along such a path by forcing the plant output information included in each state on this path. In fact, this corresponds to the procedure of the HiL Verification approach. However, the visualization of the plant simulation would be available.
The time, which is necessary to apply the verification framework of this thesis, depends on the regarded system model. According to model checking, the time for analysis exponentially raises the more trajectories have to be considered. To keep the size manageably, the closed loop of plant and controller is considered, which significantly limits the amount of state space trajectories in contrast to open-loop technologies. However, the model checking algorithm has to be implemented efficiently. Optimizing this task for the TMoC would shorten the time, which is necessary to handle larger-scale systems in a meaningful time.
HiL and SiL testing complement each other because of their specific application areas. According to verification, the interrelation of the computed dynamic graphs holds a

certain potential. For both cases, the utilized plant model is exactly the same. However, for the HiL approach the controller is just characterized through its in- and outputs and its chosen internal variables. In contrast, a complete controller model is provided to perform the SiL approach. Intuitively, the HiL dynamic graph is a sub-graph of the SiL one because the whole controller is abstracted and considered as a black box. If both verification technologies are applied, the corresponding graphs could be compared to check whether the sub-graph relation holds. Doing so, a further quality criterion is achieved to survey the correct system behavior. The concrete definitions as well as the implementation could be part of future work.
Besides the tool-related improvements of the framework, real-time considerations could further enrich the application in practice. To do so, the discrete intervals of ${}_D$TNCES could be interpreted with concrete dimensionful values. These could either be extracted from the plant simulation model or from the control code implementation. In addition, the communication delays between controller hardware and simulation or verification PC could be evaluated by measuring the retardation. The real-time information could be added to the dynamic graph to check whether a particular action is carried out in a specific time. In addition, conclusions according to the response time of the controller hardware can be drawn.
Summing up, it is the opinion of the author of this thesis that the presented framework is appropriate to include formal analysis to the plant engineering practice. Most steps have been automated to prevent the user from dull theory. Thereby, special focus has been on the application of domain-specific approaches, which should increase the acceptance. However, the framework remains flexible for prospective findings, and for this, it is arbitrarily expandable.

A. Temporal Logics

Computation Tree Logic

Figures A.1 to A.10 show the basic CTL expressions, which are provided in Definition 2.5.2.

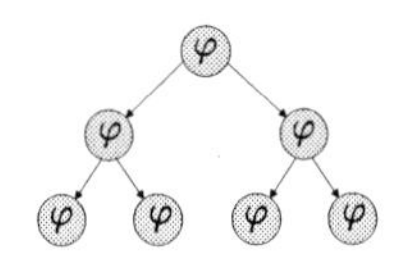

Figure A.1.: $z_0 \models AG\,\varphi$.

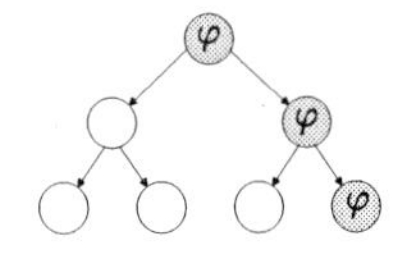

Figure A.2.: $z_0 \models EG\,\varphi$.

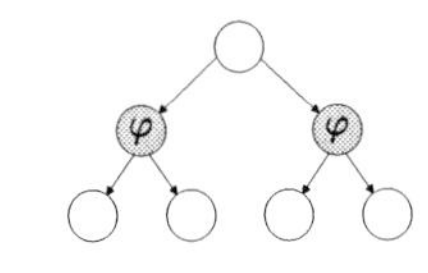

Figure A.3.: $z_0 \models AX\,\varphi$.

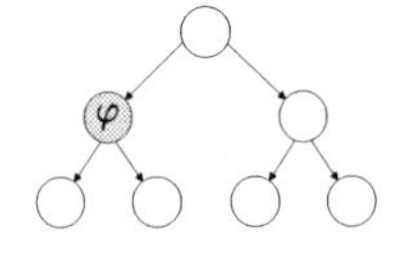

Figure A.4.: $z_0 \models EX\,\varphi$.

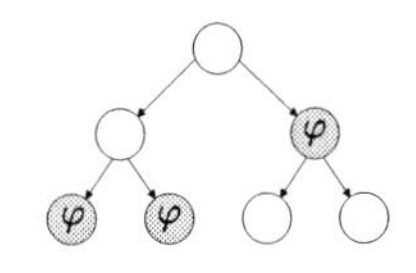

Figure A.5.: $z_0 \models AF\,\varphi$.

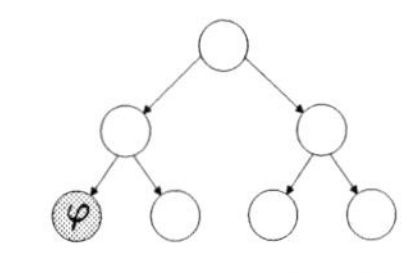

Figure A.6.: $z_0 \models EF\,\varphi$.

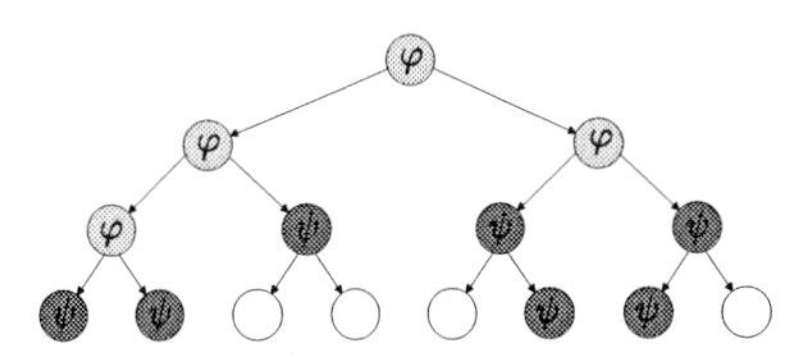

Figure A.7.: $z_0 \models A\,[\varphi\,U\,\psi]$.

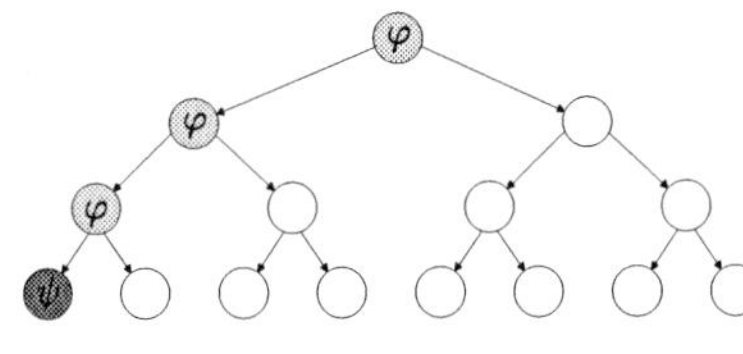

Figure A.8.: $z_0 \models E\,[\varphi\,U\,\psi]$.

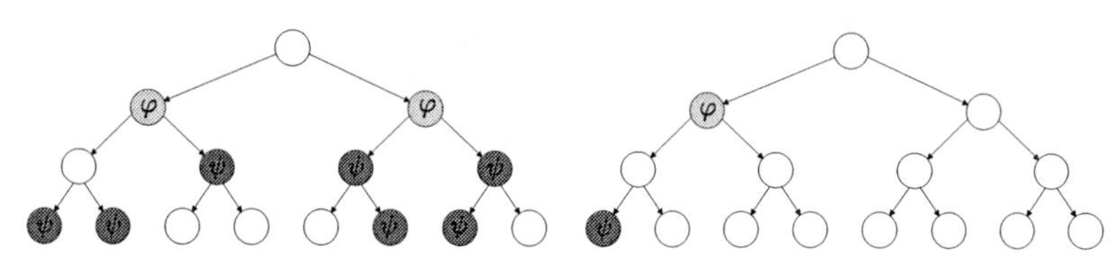

Figure A.9.: $z_0 \models A\,[\varphi\, B\, \psi]$.

Figure A.10.: $z_0 \models E\,[\varphi\, B\, \psi]$.

Extended Computation Tree Logic

Figures A.11 and A.12 show two reachability graphs and give examples for $A\,\tau\, X\,\varphi$ and $E\,\tau\, X\,\varphi$. The edges are labeled with executable steps. z_0 is the initial state. Note that the truth value of φ in state z_3 is not relevant because this state is ignored due to the transition formula t_1. For Figure A.11 it holds: $z_0 \models A\,t_1\, X\,\varphi$ and $z_0 \models E\,t_1\, X\,\varphi$. For Figure A.12 it holds: $z_0 \not\models A\,t_1\, X\,\varphi$ and $z_0 \models E\,t_1\, X\,\varphi$.

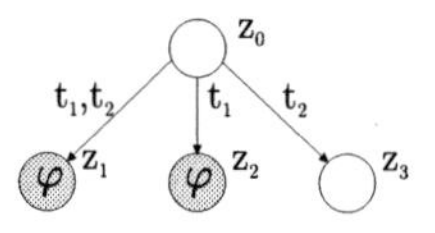

Figure A.11.: ECTL formula [SR02].

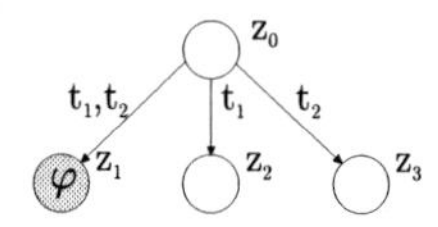

Figure A.12.: ECTL formula [SR02].

B. SOTL Grammar

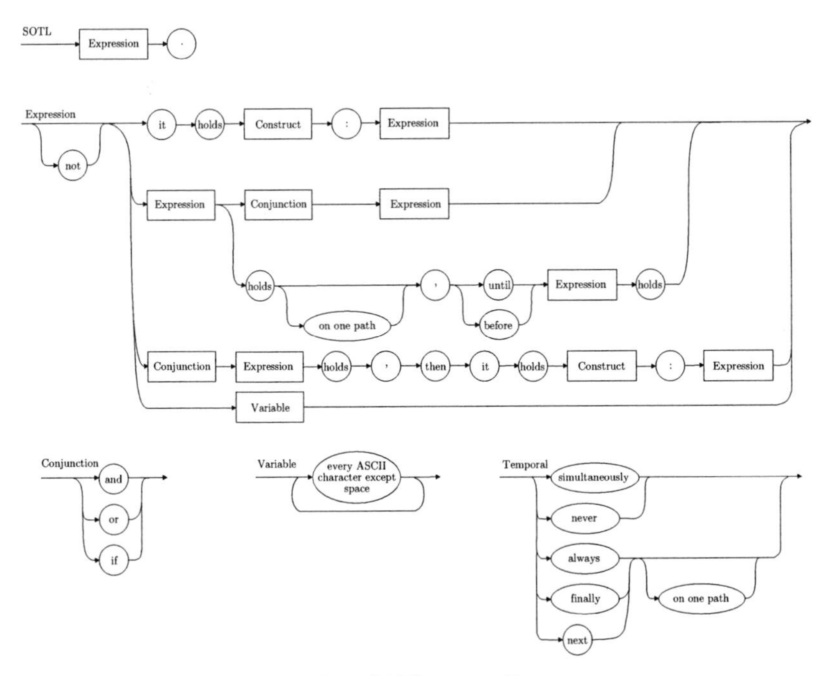

Figure B.1.: SOTL syntax diagram.

- $\mathcal{S}$ ⟦ Expression . ⟧ = $\mathcal{E}$ ⟦ Expression ⟧
- $\mathcal{E}$ ⟦ not Expression ⟧ = ! $\mathcal{E}$ ⟦ Expression ⟧
- $\mathcal{E}$ ⟦ it holds Temporal : Expression ⟧ = $\mathcal{T}$ ⟦ Temporal ⟧ $\mathcal{E}$ ⟦ Expression ⟧
- $\mathcal{E}$ ⟦ Expression_1 Conjunction Expression_2 ⟧ =
  $\mathcal{E}$ ⟦ Expression_1 ⟧ $\mathcal{C}$ ⟦ Conjunction ⟧ $\mathcal{E}$ ⟦ Expression_2 ⟧
- $\mathcal{E}$ ⟦ Expression_1 holds , until Expression_2 holds ⟧ =
  A [$\mathcal{E}$ ⟦ Expression_1 ⟧ U $\mathcal{E}$ ⟦ Expression_2 ⟧]
- $\mathcal{E}$ ⟦ Expression_1 holds , before Expression_2 holds ⟧ =
  A [$\mathcal{E}$ ⟦ Expression_1 ⟧ B $\mathcal{E}$ ⟦ Expression_2 ⟧]
- $\mathcal{E}$ ⟦ Expression_1 holds on one path , until Expression_2 holds ⟧ =
  E [$\mathcal{E}$ ⟦ Expression_1 ⟧ U $\mathcal{E}$ ⟦ Expression_2 ⟧]
- $\mathcal{E}$ ⟦ Expression_1 holds on one path , before Expression_2 holds ⟧ =
  E [$\mathcal{E}$ ⟦ Expression_1 ⟧ B $\mathcal{E}$ ⟦ Expression_2 ⟧]
- $\mathcal{E}$ ⟦ if Expression_1 holds then it holds Temporal : Expression_2 ⟧ =
  $\mathcal{E}$ ⟦ Expression_1 ⟧ -> $\mathcal{T}$ ⟦ Temporal ⟧ $\mathcal{E}$ ⟦ Expression_2 ⟧
- $\mathcal{E}$ ⟦ Variable ⟧ = Variable
- $\mathcal{C}$ ⟦ and ⟧ = &
- $\mathcal{C}$ ⟦ or ⟧ = V
- $\mathcal{T}$ ⟦ never ⟧ = ! AG
- $\mathcal{T}$ ⟦ always ⟧ = AG
- $\mathcal{T}$ ⟦ always on one path ⟧ = EG
- $\mathcal{T}$ ⟦ finally ⟧ = AF
- $\mathcal{T}$ ⟦ finally on one path ⟧ = EF
- $\mathcal{T}$ ⟦ next ⟧ = AX
- $\mathcal{T}$ ⟦ next on one path ⟧ = EX

Figure B.2.: SOTL translation rules.

C. Algorithms

Algorithm C.1: Procedure for labeling the states satisfying $EX\ \varphi$

```
1  procedure CheckEX(φ)
2    for all t ∈ Z such that ∃ (z,σ,δσ,t) ∈ ED do //consider all target states according
3                                                  //to transitions in E
4      if EX φ ∉ label(z) and φ ∈ label(t) then    //if state z is not labeled with EX φ
5                                                  //and state t is labeled with φ
6        label(z) := label(z) ∪ {EX φ};            //label state z
7      end if;
8    end for all;
9  end procedure;
```

Algorithm C.2: Procedure for labeling the states satisfying $E\,[\,\varphi\, U\, \psi\,]$ [CGP00]

```
1  procedure CheckEU(φ,ψ)
2    Z' := { z ∈ Z | ψ ∈ label(z)};        //store all states fulfilling ψ to Z'
3    for all z ∈ Z' do
4      label(z) := label(z) ∪ {E [ φ U ψ ]};                //label all states in Z'
5    while Z' ≠ ∅ do                        //as long as Z' is not empty
6      choose z ∈ Z';                       //take state of Z'
7      Z' := Z' \ {z};                      //delete z from Z'
8      for all t ∈ Z such that ∃ (t,σ,δσ,z) ∈ ED do
9                                           //consider all source states
10                                          //according to transitions in ED
11       if E [ φ U ψ ] ∉ label(t) and φ ∈ label(t) then    //if state t is not labeled
12                                                          //with E [ φ U ψ ] and
13                                                          //state t is labeled with φ
14         label(t) := label(t) ∪ {E [ φ U ψ ]};            //label state t
15         Z' := Z' ∪ {t};                  //store state t to Z'
16       end if;
17     end for all;
18   end while;
19 end procedure;
```

Algorithm C.3: Procedure for labeling the states satisfying $EG\varphi$

```
1  procedure CheckEG(φ)
2    Z' := {z ∈ Z | φ ∈ label(z)};        //store all states fulfilling φ to Z'
3    E_D' := {(z1,σ,δσ,z2) ∈ E_D | z1,z2 ∈ Z'}
4                                          //store all transitions that connect
5                                          //two states of Z' to E'
6    for all z ∈ Z' do                     //go through all states of Z'
7      U := Z';                            //store all states of Z' to U
8      S := ∅;                             //clear S
9      T := E';                            //store all transitions of E' to T
10     trajectory(z);                      //call procedure recursively with current state
11   end for all;
12 end procedure;
13
14 procedure trajectory(z)
15   S.push(z);                            //store z to stack S
16   U := U \ {z};                         //remove z from U
17   if T = ∅ then return;                 //terminate if transition set is empty
18   for all (z,σ,δσ,z') in T do           //go through all transitions of T
19     if z' ∈ U then                      //target state is in U
20       T := T \ {(z,σ,δσ,z')};           //remove transition from T
21       trajectory(z');                   //call procedure recursively with new state
22       return;                           //terminate
23     end if;
24     if z' ∈ S then                      //target state is in S
25                                         //so it has already been considered
26       label(z[S[0]]) := label(z[S[0]]) ∪ {EG φ}  //label root state
27       return;                           //terminate
28     end if;
29   end for all;
30   if T ≠ ∅ then                         //transition set is not empty
31     S.pop;                              //delete last state from S
32     U[0] := S.pop;                      //store last state of S to first position of U
33                                         //and delete last state from S
34     trajectory(U[0]);                   //call procedure recursively with new state
35     return;                             //terminate
36   end if;
37 end procedure;
```

Algorithm C.4: Algorithm in pseudocode for HiL state space computation

```
1  procedure HiLSimulation()
2    statespace.clear();                   //initialize state space
3    currentstate := create_init_state(); //create initial state considering the labels
4                                          //and clocks of all places in the imported
5                                          //model
6    statespace.add(currentstate);         //add state to state space
7    complete := false;                    //set exit condition to false
8    while ! complete do
9      ComputeStateSpace();                //call calculation function
10   end while;
11 end procedure;
12
13 procedure ComputeStateSpace()
14   outputs := read_output_states();      //get all possible controller output states,
15                                         //which were recorded for a predefined time,
16                                         //(e.g. the maximum of all timers in the
17                                         //PLC code)
18   for i=0;i<outputs.size();i++ do       //iterate over all recorded configurations
19                                         //(i.e. consider all controller output states)
20     currentstate.create(currentstate,outputs[i]);//create new currentstate by
21                                              //including output information to
22                                              //existing currentstate
23     if statespace.contains(currentstate) then    //new state already existing
24       continue;                                  //continue loop
25     end if;
26     else                                //new state does not exist
27       statespace.add(currentstate);     //add new state to state space
28       compute_all(currentstate);        //calculate all successor states i+n starting
29                                         //from currentstate i
30       mark_branch_states();             //mark all states, which contain new plant
31                                         //output (controller input) information
32     end else;
33   end for;
34   currentstate := get_branch_state(statespace);  //iterate over states and get first,
35                                                  //which is labeled as branchstate
36   if currentstate.empty() then                   //state space complete
37     complete := true;
38     return;
39   end if;
40   else
41     drive_controller_to_branch_state();          //get path from initial state to
42                                                  //branchstate and drive controller
43                                                  //state by state to this branchstate
44     write_input_states();                        //update PLC inputs
45   end else;
46 end procedure;
```

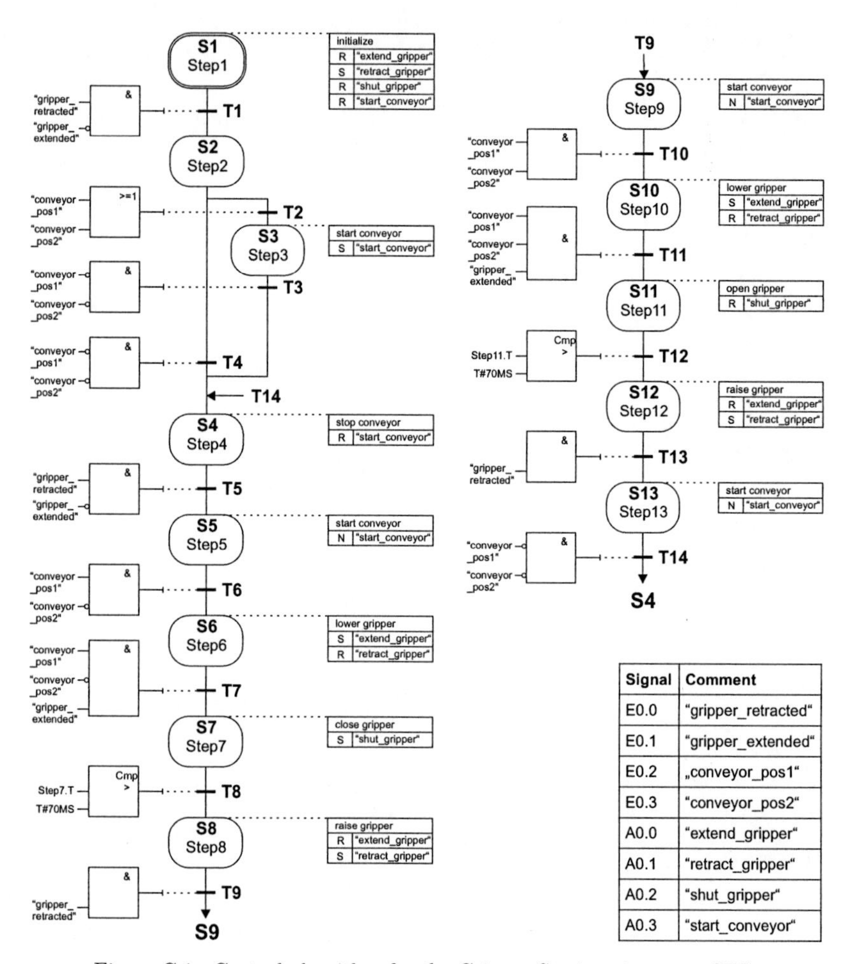

Signal	Comment
E0.0	"gripper_retracted"
E0.1	"gripper_extended"
E0.2	„conveyor_pos1"
E0.3	"conveyor_pos2"
A0.0	"extend_gripper"
A0.1	"retract_gripper"
A0.2	"shut_gripper"
A0.3	"start_conveyor"

Figure C.1.: Control algorithm for the Gripper Station given as a SFC.

D. Example for Model Checking

Figure D.1 shows an example for an $_D$TNCES consisting of 4 $_D$TNCE Modules. The $_D$TNCES is safe as the places have the capacity of 1. The modules are interconnected through condition, inhibitor and event arcs.

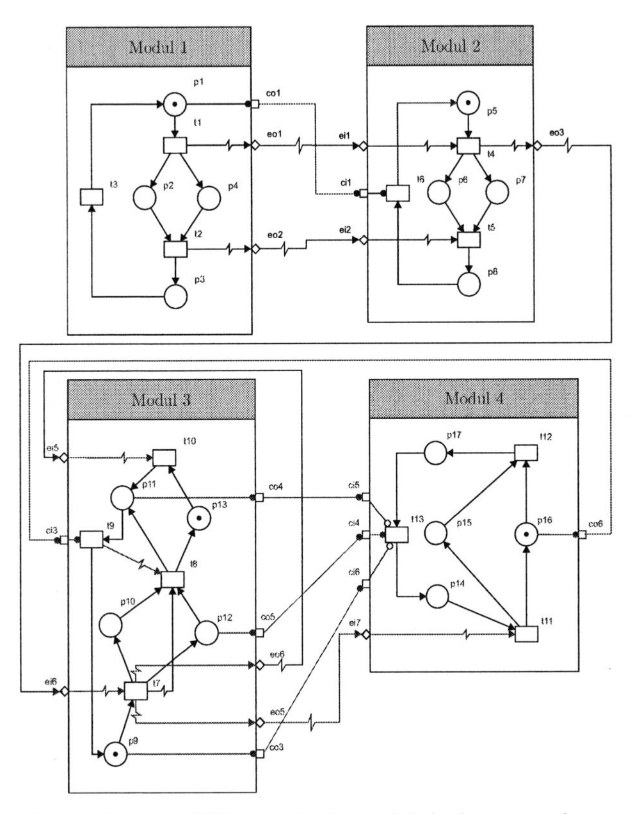

Figure D.1.: $_D$TNCES example for model checking procedure.

The reachability graph that was computed for the $_D$TNCES in Figure D.1 consists of 11 states and is given in Figure D.2.

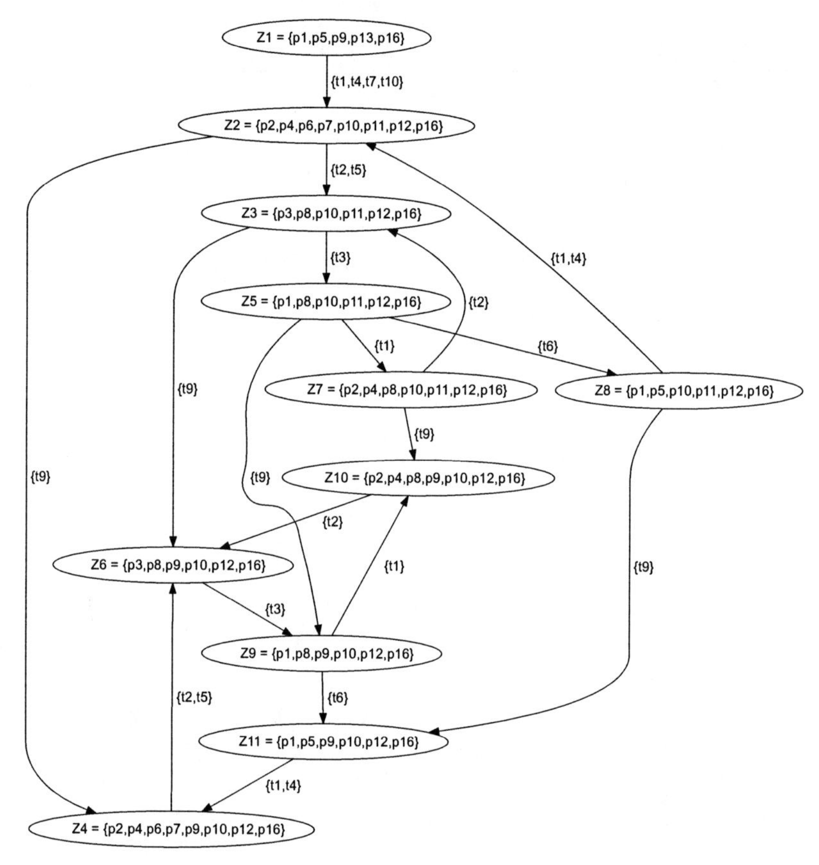

Figure D.2.: Calculated reachability graph of the $_D$TNCES in Figure D.1.

To show the application of the model checking algorithm, two properties are specified. For some reason, p3 and p7 must not be marked at the same time. This is expressed by the formula $AG\,!\,(\,p3 \wedge p7\,)$. Another property may be that it should always hold if p1 is marked, then p2 will be marked in the very next state. This is expressed by the formula $AG\,(p1 \rightarrow AX\,p2\,)$. According to Definition 2.5.3, the two formulas can be rewritten as given below.

$$AG\,!\,(p3 \wedge p7) \quad \equiv \quad !\,E\,[\,true\,U\,(\,p3 \wedge p7\,)\,] \tag{D.1}$$
$$AG\,(p1 \rightarrow AX\,p2\,) \quad \equiv \quad !\,E\,[\,true\,U\,!\,(\,!\,p1 \vee !\,EX\,!\,p2\,)\,] \tag{D.2}$$

The model checking algorithm processes formula D.1 by first considering the most nested subformulas and then working outward. Doing so, the set of fulfilling states is determined as follows.

$$
\begin{aligned}
&Z(p3) &&= &&\{3,6\}\\
&Z(p7) &&= &&\{2,4\}\\
&Z(p3\ \wedge\ p7) &&= &&\emptyset\\
&Z(E\ [\ true\ U\ (\ p3\ \wedge\ p7\ )\ ]) &&= &&\emptyset\\
&Z(!\ E\ [\ true\ U\ (\ p3\ \wedge\ p7\ )\ ]) &&= &&\{1,2,3,4,5,6,7,8,9,10,11\}
\end{aligned}
$$

Since all states are in the set, the formula holds true. Analogously, formula D.2 is processed as follows.

$$
\begin{aligned}
&Z(p2) &&= &&\{2,4,7,10\}\\
&Z(!\ p2) &&= &&\{1,3,5,6,8,9,11\}\\
&Z(EX\ !\ p2) &&= &&\{2,3,4,5,6,7,8,9,10\}\\
&Z(!\ EX\ !\ p2) &&= &&\{1,11\}\\
&Z(p1) &&= &&\{1,5,8,9,11\}\\
&Z(!\ p1) &&= &&\{2,3,4,6,7,10\}\\
&Z(!\ p1 \vee !\ EX\ !\ p2) &&= &&\{1,2,3,4,6,7,10,11\}\\
&Z(!\ (\ !\ p1 \vee !\ EX\ !\ p2\ )) &&= &&\{5,8,9\}\\
&Z(E\ [\ true\ U\ !\ (\ !\ p1 \vee !\ EX\ !\ p2\ )\ ]) &&= &&\{1,2,3,4,5,6,7,8,9,10,11\}\\
&Z(!\ E\ [\ true\ U\ !\ (\ !\ p1 \vee !\ EX\ !\ p2\ )\ ]) &&= &&\emptyset
\end{aligned}
$$

The set of states fulfilling formula D.2 is empty, and for this, the formula holds false. A trajectory violating the formula is for example given by the sequence of states $\{3,5,8,2\}$ because $p2$ is not marked in state 8 after $p1$ was marked in state 5.

Bibliography

[AH74] A. V. Aho and J. E. Hopcroft. *The Design and Analysis of Computer Algorithms.* Addison-Wesley Longman Publishing Co., Inc., Boston, MA, USA, 1. edition, 1974.

[Ain96] P. Ainsworth. The millennium bug [computer systems]. *IEE Review*, 42(4):140–142, July 1996.

[Aut12] AutomationML Whitepaper Part 1 - 4. `www.automationml.org`, 2012.

[BAPM83] M. Ben-Ari, A. Pnueli, and Z. Manna. The Temporal Logic of Branching Time. *Acta Informatica*, 20(3):207–226, 1983.

[BK08] C. Baier and J.-P. Katoen. *Principles of Model Checking.* The MIT Press, Cambridge, MA, USA, 1. edition, 2008.

[CES86] E. M. Clarke, E. A. Emerson, and A. P. Sistla. Automatic Verification of Finite-State Concurrent Systems Using Temporal Logic Specifications. *ACM Trans. Program. Lang. Syst.*, 8:244–263, April 1986.

[CF05] H. Chockler and K. Fisler. Temporal Modalities for Concisely Capturing Timing Diagrams. *Correct Hardware Design and Verification Methods, Lecture Notes in Computer Science (LNCS)*, 3725:176–190, 2005.

[CGP00] E. M. Clarke, O. Grumberg, and D. A. Peled. *Model Checking.* MIT Press, Cambridge, MA, USA, 2000.

[DIN05] DIN EN ISO/IEC 17025: General requirements for the competence of testing and calibration laboratories, August 2005.

[GDD09] D. Gasevic, D. Djuric, and V. Devedzic. *Model Driven Engineering and Ontology Development.* Springer-Verlag, 2. edition, 2009.

[Ger11] C. Gerber. *Implementation and Verification of Distributed Control Systems*, volume 7 of *Hallenser Schriften zur Automatisierungstechnik.* Logos Verlag GmbH, Berlin, Germany, 2011.

[GPH10] C. Gerber, S. Preuße, and H.-M. Hanisch. A Complete Framework for Controller Verification in Manufacturing. In *15th IEEE International Conference on Emerging Technologies and Factory Automation (ETFA)*, Bilbao, Spain, September 2010. IEEE. Index: MF-001279.

[Han98] H.-M. Hanisch. Petri-Netze im geschlossenen Kreis: Stand und Tendenzen. In D. Abel and K. Lemmer, editors, *Theorie ereignisdiskreter Systeme*, pages 51–72. Oldenbourg-Verlag, München, Germany, 1998. In German.

[HC95] H.-M. Hanisch and U. Christmann. Modeling and Analysis of a Polymer Production Plant by Means of Arc-Timed Petri Nets. *International Journal of Flexible Automation and Integrated Manufacturing*, 3(1):33–46, 1995.

[HCG+12] T. Holm, L. Christiansen, M. Göring, T. Jäger, and A. Fay. ISO 15926 vs. IEC 62424 - Comparison of Plant Structure Modeling Concepts. In *17th IEEE International Conference on Emerging Technologies and Factory Automation (ETFA)*, pages 1–8, Kraków, Poland, September 2012. IEEE. Index: 158.

[Hir10] M. Hirsch. *Systematic Design of Distributed Industrial Manufacturing Control Systems*, volume 6 of *Hallenser Schriften zur Automatisierungstechnik*. Logos Verlag GmbH, Berlin, Germany, 2010.

[HMD01] M. Heiner, T. Mertke, and P. Deussen. A Safety-Oriented Technical Language for the Requirement Specification in Control Engineering. In *Computer Science Reports 09/01*, page 65ff, BTU Cottbus, May 2001.

[Hol97] G. J. Holzmann. The Model Checker SPIN. *IEEE Transactions on Software Engineering (TSE)*, 23(5):279–295, May 1997.

[HTLW97] H.-M. Hanisch, J. Thieme, A. Lüder, and O. Wienhold. Modeling of PLC Behavior by Means of Timed Net Condition/Event Systems. In *6th IEEE International Conference on Emerging Technologies and Factory Automation (ETFA)*, pages 391–396, Los Angeles, USA, September 1997. IEEE Press.

[IEC03] IEC 61131: Programmable controllers - Part 3: Programming languages, January 2003.

[IEC04] ISO/IEC 15909-1: High-level Petri Nets - Concepts, Definitions and Graphical Notation., 2004.

[IEC05] IEC 61499: Function blocks - Part 1: Architecture, January 2005.

[IEC08a] IEC 62424: Specification for Representation of process control engineering requests in P&I Diagrams and for data exchange between P&ID tools and PCE-CAE, 2008.

[IEC08b] ISO/IEC 9001: Quality management systems – Requirements, December 2008.

[IEC10] IEC 61508: Functional Safety, April 2010.

[ISO96] ISO/IEC 14977: Information technology – Syntactic metalanguage – Extended BNF, 1996.

[ISO04] ISO 15926-1: Industrial automation systems and integration - Integration of life-cycle data for process plants including oil and gas production facilities - Part 1: Overview and fundamental principles, December 2004.

[ISO09] ISO 16100-1: Industrial automation systems and integration – Manufacturing software capability profiling for interoperability – Part 1: Framework, December 2009.

[ISO10] ISO 16100-1: Specification for diagrams for process industry – Part 1: General rules, February 2010.

[ISO11] ISO 26262-1: Road vehicles – Functional safety – Part 1: Vocabulary, November 2011.

[Joh07] T. L. Johnson. Improving automation software dependability: A role for formal methods? *Control Engineering Practice*, 15:1403–1415, 2007.

[Kar09] S. Karras. *Systematischer modellgestützter Entwurf von Steuerungen für Fertigungssysteme*, volume 4 of *Hallenser Schriften zur Automatisierungstechnik*. Logos Verlag GmbH, Berlin, Germany, 2009. In German.

[KS95] F. Korf and R. Schlör. Symbolic Timing Diagrams. In *Formal Development of Reactive Systems - Case Study Production Cell*, pages 311–331, London, UK, 1995. Springer-Verlag.

[LCL87] F. J. Lin, P. M. Chu, and M. T. Liu. Protocol verification using reachability analysis: the state space explosion problem and relief strategies. *SIGCOMM Comput. Commun. Rev.*, 17(5):126–135, August 1987.

[McM93] K. L. McMillan. *Symbolic Model Checking*. Kluwer Academic Publishers, Norwell, MA, USA, 1. edition, 1993.

[MHH07] D. Missal, M. Hirsch, and H.-M. Hanisch. Hierarchical Distributed Controllers - Design and Verification. In *12th IEEE International Conference on Emerging Technologies and Factory Automation (ETFA)*, pages 657–664, Patras, Greece, 2007.

[Mis12] D. Missal. *Formal Synthesis of Safety Controller Code for Distributed Controllers*, volume 9 of *Hallenser Schriften zur Automatisierungstechnik*. Logos Verlag GmbH, Berlin, Germany, 2012.

[PGH10] S. Preuße, C. Gerber, and H.-M. Hanisch. Design Approaches for IEC 61499 Control Applications. In *INFORMATIK 2010 - 40. Jahrestagung der Gesellschaft für Informatik e.V. (GI), Lecture Notes in Informatics (LNI)*, pages 469–479, Leipzig, Germany, September 2010.

[PH11] S. Preuße and H.-M. Hanisch. Verifying Functional and Non-Functional Properties of Manufacturing Control Systems. In *3rd International Workshop on Dependable Control of Discrete Systems (DCDS)*, pages 41–46, Saarbrücken, Germany, June 2011. IEEE.

[PLH12] S. Preuße, H.-C. Lapp, and H.-M. Hanisch. Closed-loop System Modeling, Validation, and Verification. In *17th IEEE International Conference on Emerging Technologies and Factory Automation (ETFA)*, pages 1–8, Kraków, Poland, September 2012. IEEE. Index: 99.

[PMG+11] S. Preuße, D. Missal, C. Gerber, M. Hirsch, and H.-M. Hanisch. On the Use of Model-Based IEC 61499 Controller Design. *Int. J. of Discrete Event Control Systems (IJDECS)*, 1(1):115–128, January 2011.

[Pnu77] A. Pnueli. The temporal logic of programs. In *Symposium on Foundations of Computer Science (FOCS)*, volume 18, pages 46–57. IEEE Computer Society, 1977.

[QS82] J.-P. Queille and J. Sifakis. Specification and Verification of Concurrent Systems in CESAR. In *5th Colloquium on International Symposium on Programming (ISP)*, volume 137 of *Lecture Notes in Computer Science*, pages 337–351, Turin, Italy, April 1982. Springer-Verlag.

[Sch01] R.C. Schlör. *Symbolic Timing Diagrams: A Visual Formalism for Model Verification.* PhD thesis, Carl-von-Ossietzky University of Oldenburg, Germany, 2001.

[Sch04] K. Schneider. *Verification of Reactive Systems: Formal Methods and Algorithms.* Texts in Theoretical Computer Science. An EATCS Series. Springer-Verlag, Berlin, Germany, December 2004.

[Sch08] S. Schewe. *Synthesis of Distributed Systems.* PhD thesis, Naturwissenschaftlich-Technische Fakultäten der Universität des Saarlandes, Saarbrücken, 2008.

[SD93] R.C. Schlör and W. Damm. Specification and Verification of System-Level Hardware Designs using Timing Diagrams. In *European Conference on Design Automation (EDAC)*, pages 518–524, Paris, France, 1993. IEEE Computer Society Press.

[SE12] A. Schüller and U. Epple. PandIX - Exchanging P&I diagram model data. In *17th IEEE International Conference on Emerging Technologies and Factory Automation (ETFA)*, pages 1–8, Kraków, Poland, September 2012. IEEE. Index: 31.

[SJW98] R. Schlör, B. Josko, and D. Werth. Using a visual formalism for design verification in industrial environments. In *VISUAL'98, Lecture Notes in Computer Science*, volume 1385, pages 208–221. Springer-Verlag, 1998.

[SR02] P.H. Starke and S. Roch. Analysing Signal-Net Systems. *Informatikberichte, Humboldt-Universität zu Berlin*, 162, September 2002.

[TBdS07] E. Tisserant, L. Bessard, and M. d. Sousa. An Open Source IEC 61131-3 Integrated Development Environment. In *5th IEEE International Confer-*

ence on Industrial Informatics (INDIN), pages 183–187, Vienna, Austia, July 2007.

[Tri09] S. Tripakis. Checking Timed Büchi Automata Emptiness on Simulation Graphs. *ACM Transactions on Computational Logic (ACM T COMPUT LOG)*, 10(3):1–19, April 2009.

[VDI96] VDI 3633: Simulation of systems in materials handling, logistics and production - Clarification of terms, November 1996.

[VHPY09] V. Vyatkin, H.-M. Hanisch, C. Pang, and J. Yang. Application of Closed-Loop Modelling in Integrated Component Design and Validation of Manufacturing Automation. *IEEE Transactions on Systems, Machine and Cybernetics (SMC)*, 39(1):17–28, 2009.

[Wit05] H. Wittke. *An Environment for Compositional Specification Verification of Complex Embedded Systems.* PhD thesis, Carl-von-Ossietzky University of Oldenburg, Gemany, 2005.

[WLH11] T. Winkler, H.-C. Lapp, and H.-M. Hanisch. A new Model Structure based Synthesis Approach for Distributed Discrete Process Control. In *9th IEEE International Conference on Industrial Informatics (INDIN)*, pages 527–532, Caparica, Lisbon, Portugal, July 2011.

Index

Curriculum Vitae

	Personal Details
Name	Sebastian Preuße
Date of Birth	24th April 1983
Place of Birth	Bernburg, Germany
	Professional Experience
06/08 – 05/14	Scientific Assistant, Martin-Luther-University, Institute of Computer Science, Chair for Automation Technology, Halle (Saale), Germany
04/07 – 09/07	Internship, LAE Engineering GmbH, Nußloch, Germany
04/06 – 03/07	Student Assistant, Martin-Luther-University, Institute of Computer Science, Chair for Automation Technology, Halle (Saale), Germany
07/04 – 08/04	Internship, Solvay GmbH, Bernburg, Germany
	Education
10/02 – 05/08	Martin-Luther-University, Halle (Saale), Germany Subject of Studies: Informatics in Engineering Major Subject: Automation Technology Minor Subjects: Communication Technology, Computer Science, Process Engineering Degree: Diplom-Ingenieur (equivalent to Master's degree)
1993 – 2002	Grammar School, Hermann-Hellriegel-Gymnasium, Bernburg, Germany Graduation: Abitur (equivalent to A level)
1989 – 1993	Elementary School, Grundschule Süd-West, Bernburg, Germany
	Research Project
06/08 – 07/11	On-The-Fly-Migration and Instant-Start-Up of Automated Systems (OMSIS), funded by the Bundesministerium für Wirtschaft und Technologie (BMWi), Promotion Code: 16 IN 0651